ELECTRICIAN'S
VEST POCKET
REFERENCE BOOK

Other Books in the Prentice-Hall Vest Pocket Reference Book Series

BUILDER'S VEST POCKET REFERENCE BOOK
by William Hornung

ELECTRONICS VEST POCKET REFERENCE BOOK
by Harry Thomas

MECHANIC'S VEST POCKET REFERENCE BOOK
by J. E. Wolfe and E. R. Phelps

PIPEFITTER'S AND PLUMBER'S VEST POCKET REFERENCE BOOK
by George K. Bachmann

ELECTRICIAN'S VEST POCKET REFERENCE BOOK

Revised Edition

by

HENRY B. HANSTEEN

PRENTICE-HALL, INC.
Englewood Cliffs, N.J.

*Electrician's Vest Pocket Reference Book
—Revised Edition* by Henry B. Hansteen

Printed in the United States of America
Prentice-Hall International, Inc., London
Prentice-Hall of Australia, Pty. Ltd., North Sydney
Prentice-Hall of Canada, Ltd., Toronto
Prentice-Hall of India Private Ltd., New Delhi
Prentice-Hall of Japan, Inc., Tokyo

Library of Congress Cataloging in Publication Data

Hansteen, Henry B
Electrician's vest pocket reference book.

1. Electric wiring—Tables, calculations, etc.
I. Title.
TK3205.H3 1973 621.319'24 72-12321
ISBN 0-13-247957-5

10 9 8 7 6

TABLE OF CONTENTS

Table		Page
24	Maximum Rating or Setting of Motor Branch-Circuit Protective Devices	86
25	Demand Loads for Household Appliances	90
26	Voltage Drop Data	94
27	American Wire Gauge	98
28	Kilowatts to Amperes Conversion Tables	99
29	Horsepower to Amperes Conversion Tables	101
30	Recommended Bus Spacings for Different Voltages	104
31	Sizes, Weights and Current Carrying Capacity of Drawn Rectangular Copper Bars	105
32	Useful Information on Rigid Steel Conduit	107
33	Weights of Cabinets and Boxes	113
34	Weights of A-C Motors	115
35	Weights of D-C Motors	115
36	Transformer Weights	115
37	Typical Data on Light Sources	116

INTRODUCTION

The Electrician's Vest Pocket Reference Book is a companion volume to the Mechanic's Reference Book and the Builder's Reference Book. It contains all the important Tables of the latest (1971) National Electric Code, as well as other related information which should prove useful to the practicing electrician on the job.

In order that the Tables abstracted from the NEC (National Electric Code) may be identified with the Code itself and related material, the NEC Table numbers are included in parentheses following the sequential numbers of the Tables used for indexing. All Section references are to the 1971 NEC.

All of the NEC material is reproduced with the permission of the National Board of Fire Underwriters. Other tabular data is included through the courtesy of the American Iron and Steel Institute and the National Electrical Contractors Association.

ELECTRICIAN'S VEST POCKET REFERENCE BOOK

TABLE 1 (310-2a)—Conductor Application by Type of Insulation and Maximum Operating Temperature

Trade Name	Type Letter	Max. Operating Temp.	Application Provisions
Rubber-Covered Fixture Wire Solid or 7-Strand	*RF-1	60°C 140°F	Fixture wiring. Limited to 300 volts.
	*RF-2	60°C 140°F	Fixture wiring, and as permitted in Section 725-14.
Rubbered-Covered Fixture Wire Flexible Stranding	*FF-1	60°C 140°F	Fixture wiring. Limited to 300 volts.
	*FF-2	60°C 140°F	Fixture wiring, and as permitted in Section 725-14.

Heat-Resistant Rubber-Covered Fixture Wire Solid or 7-Strand	*RFH-1	75°C 167°F	Fixture wiring. Limited to 300 volts.
	*RFH-2	75°C 167°F	Fixture wiring, and as permitted in Section 725-14.
Heat-Resistant Rubber-Covered Fixture Wire Flexible Stranding	*FFH-1	75°C 167°F	Fixture wiring. Limited to 300 volts.
	*FFH-2	75°C 167°F	Fixture wiring, and as permitted in Section 725-14.
Thermoplastic-Covered Fixture Wire—Solid or Stranded	*TF	60°C 140°F	Fixture wiring, and as permitted in Section 725-14.

*Fixture wires are not intended for installation as branch-circuit conductors except as permitted in Section 725-14.

TABLE 1 (310-2a) (*Cont.*)

Trade Name	Type Letter	Max. Operating Temp.	Application Provisions
Thermoplastic-Covered Fixture Wire—Flexible Stranding	*TFF	60°C 140°F	Fixture wiring, and as permitted in Section 725-14.
Heat Resistant, Thermoplastic—Covered Fixture Wire—Solid or Stranded	*TFN	90°C	Fixture wiring, and as permitted in Section 725-14.

Heat Resistant Thermoplastic—Covered Fixture Wire—Flexible Stranding	*TFFN	90°C	Fixture wiring, and as permitted in Section 725-14.
Cotton-Covered, Heat-Resistant, Fixture Wire	*CF	90°C 194°F	Fixture wiring. Limited to 300 volts.
Asbestos-Covered Heat-Resistant, Fixture Wire	*AF	150°C 302°F	Fixture wiring. Limited to 300 volts, and indoor dry location.
Fluorinated Ethylene Propylene Fixture Wire Solid or 7 Strand	*PF *PGF	200°C 392°F	Fixture wiring and as permitted in Section 725-14.

*Fixture wires are not intended for installation as branch-circuit conductors except as permitted in Section 725-14.

TABLE 1 (310-2a) (*Cont.*)

Trade Name	Type Letter	Max. Operating Temp.	Application Provisions
Fluorinated Ethylene Propylene Fixture Wire	*PFF *PGFF	150°C 302°F	Fixture wiring and as permitted in Section 725-14.
Extruded Polytetra-fluoroethylene (PTFE) Solid or 7-Strand	*PTF	250°C 482°F	Fixture wire, and as permitted in Section 725-14. (Nickel or nickel-coated copper)

Extruded Polytetra-fluoroethylene (PTFE) Flexible Stranding (#26-#36 AWG)	*PTFF	150°C 302°F	Fixture wire, and as permitted in Section 725-14. (Silver or nickel-coated copper)
Silicone Rubber Insulated Fixture Wire	*SF-1	200°C 392°F	Fixture wiring. Limited to 300 volts.
Solid or 7 Strand	*SF-2	200°C 392°F	Fixture wiring and as permitted in Section 725-14.
Silicone Rubber Insulated Fixture Wire	*SFF-1	150°C 302°F	Fixture wiring. Limited to 300 volts.

*Fixture wires are not intended for installation as branch-circuit conductors except as permitted in Section 725-14.

TABLE 1 (310-2a) (*Cont.*)

Trade Name	Type Letter	Max. Operating Temp.	Application Provisions
Flexible Stranding	*SFF-2	150°C 302°F	Fixture wiring and as permitted in Section 725-14.
Heat-Resistant Rubber	RH	75°C 167°F	Dry locations.
Heat-Resistant Rubber	RHH	90°C 194°F	Dry locations.
Moisture and Heat-Resistant Rubber	RHW	75°C 167°F	Dry and wet locations. For over 2,000 volts, insulation shall be ozone-resistant.

Heat-Resistant Latex Rubber	RUH	75°C	Dry locations.
Moisture Resistant Latex Rubber	RUW	60°C 140°F	Dry and wet locations.
Thermoplastic	T	60°C 140°F	Dry locations.
Moisture-Resistant Thermoplastic	TW	60°C 140°F	Dry and wet locations.
Heat-Resistant Thermoplastic	THHN	90°C 194°F	Dry locations.

*Fixture wires are not intended for installation as branch-circuit conductors except as permitted in Section 725-14.

TABLE 1 (310-2a) (*Cont.*)

Trade Name	Type Letter	Max. Operating Temp.	Application Provisions
Moisture and Heat-Resistant Thermoplastic	THW	75°C 167°F	Dry and wet locations.
		90°C 194°F	Special applications *within* electric discharge lighting equipment. Limited to 1000 open-circuit volts or less. (Size 14-8 only as permitted in Section 410-26.)
Moisture and Heat-Resistant Thermoplastic	THWN	75°C 167°F	Dry and wet locations.

Moisture and Heat-Resistant Cross-Linked Synthetic Polymer	XHHW	90°C 194°F	Dry locations.
		75°C 167°F	Wet locations.
Moisture-, Heat-, and Oil-Resistant Thermoplastic	MTW	60°C 140°F	Machine Tool Wiring in wet locations as permitted in NFPA Standard No. 79 (See Article 670).
		90°C 194°F	Machine Tool Wiring in dry locations as permitted in NFPA Standard No. 79 (See Article 670).
Thermoplastic and Asbestos	TA	90°C 194°F	Switchboard wiring only.

TABLE 1 (310-2a) (*Cont.*)

Trade Name	Type Letter	Max. Operating Temp.	Application Provisions
Thermoplastic and Fibrous Outer Braid	TBS	90°C 194°F	Switchboard wiring only.
Synthetic Heat-Resistant	SIS	90°C 194°F	Switchboard wiring only.
Mineral Insulation (Metal Sheathed)	MI	85°C 185°F	Dry and wet locations.
		250°C 482°F	For special application.

Extruded Polytetra-fluoroethylene	TFE	250°C 482°F	Dry locations only. Only for leads within apparatus or within raceways connected to apparatus, or as open wiring. (Nickel or nickel-coated copper only.)
Silicone-Asbestos	SA	90°C 194°F	Dry locations.
		125°C 257°F	For special application.
Fluorinated Ethylene Propylene	FEP or FEPB	90°C 194°F	Dry locations.
		200°C 392°F	Dry locations—special applications.
Varnished Cambric	V	85°C 185°F	Dry locations only. Smaller than No. 6 by special permission.

TABLE 1 (310-2a) (*Cont.*)

Trade Name	Type Letter	Max. Operating Temp.	Application Provisions
Asbestos and Varnished Cambric	AVA	110°C 230°F	Dry locations only.
Asbestos and Varnished Cambric	AVL	110°C 230°F	Dry and wet locations.
Asbestos and Varnished Cambric	AVB	90°C 194°F	Dry locations only.

Asbestos	A	200°C 392°F	Dry locations only. Only for leads within apparatus or within raceways connected to apparatus. Limited to 300 volts.
Asbestos	AA	200°C 392°F	Dry locations only. Only for leads within apparatus or within raceways connected to apparatus or as open wiring. Limited to 300 volts.
Asbestos	AI	125°C 257°F	Dry locations only. Only for leads within apparatus or within raceways connected to apparatus. Limited to 300 volts.
Asbestos	AIA	125°C 257°F	Dry locations only. Only for leads within apparatus or within raceways connected to apparatus or as open wiring.

TABLE 1 (310-2a) *(concluded)*

Trade Name	Type Letter	Max. Operating Temp.	Application Provisions
Paper		85°C 185°F	For underground service conductors, or by special permission.

TABLE 2 (310-12)—Allowable Ampacities of Insulated Copper Conductors

This Table gives the allowable ampacities for not more than 3 conductors in a raceway or cable. When this number exceeds 3, the allowable ampacity of each conductor shall be reduced as follows:

4 to 6 conductors . . . . . 80 percent of value shown in Table
7 to 24 conductors . . . . . 70 percent of value shown in Table
25 to 42 conductors . . . . . 60 percent of value shown in Table
above 42 conductors . . . . . 50 percent of value shown in Table

Not More than Three Conductors in Raceway or Cable or
Direct Burial (Based on Ambient Temperature of 30° C. 86° F.)

Table 2 (310-12) *(Cont.)*

Size	Temperature Rating of Conductor. See Table 1 (310-2a)							
AWG MCM	60°C (140°F)	75°C (167°F)	85°C (185°F)	90°C (194°F)	110°C (230°F)	125°C (257°F)	200°C (392°F)	250°C (482°F)
	TYPES RUW (14-2), T, TW	TYPES RH, RHW, RUH (14-2), THW, THWN, XHHW	TYPES V, MI	TYPES TA, TBS, SA, AVB, SIS, FEP, FEPB, RHH, THHN, XHHW**	TYPES AVA, AVL	TYPES AI (14-8), AIA	TYPES A (14-8), AA, FEP* FEPB*	TYPE TFE (Nickel or nickel-coated copper only)
14	15	15	25	25†	30	30	30	40
12	20	20	30	30†	35	40	40	55
10	30	30	40	40†	45	50	55	75
8	40	45	50	50	60	65	70	95

6	55	65	70	70	80	85	95	120
***4	70	85	90	90	105	115	120	145
***3	80	100	105	105	120	130	145	170
***2	95	115	120	120	135	145	165	195
***1	110	130	140	140	160	170	190	220
***0	125	150	155	155	190	200	225	250
***00	145	175	185	185	215	230	250	280
000	165	200	210	210	245	265	285	315
0000	195	230	235	235	275	310	340	370

*Special use only. See Table 1 (310-2a).

**For dry locations only. See Table 1 (310-2a).

These ampacities relate only to conductors described in Table 310-2(a).

***For three-wire, single-phase residential services, the allowable ampacity of RH, RHH, RHW, THW and XHHW copper conductors shall be for sizes No. 4-100 Amp., No. 3-110 Amp., No. 2-125 Amp., No. 1-150 Amp., No. 1/0-175 Amp., and No. 2/0-200 Amp.

†The ampacities for Types FEP, FEPB, RHH, THHN, and XHHW conductors for sizes AWG 14, 12 and 10 shall be the same as designated for 75°C conductors in this Table.

For ambient temperatures over 30°C, see Correction Factors, Table 6.

TABLE 2 (310-12) *(Cont.)*

Size	Temperature Rating of Conductor. See Table 1 (310-2a)							
AWG MCM	60°C (140°F)	75°C (167°F)	85°C (185°F)	90°C (194°F)	110°C (230°F)	125°C (257°F)	200°C (392°F)	250°C (482°F)
	TYPES RUW (14-2), T, TW	TYPES RH, RHW, RUH (14-2), THW, THWN, XHHW	TYPES V, MI	TYPES TA, TBS, SA, AVB, SIS, FEP, FEPB, RHH, THHN, XHHW**	TYPES AVA, AVL	TYPES AI (14-8), AIA	TYPES A (14-8), AA, FEP* FEPB*	TYPE TFE (Nickel or nickel-coated copper only)
250	215	255	270	270	315	335	....	....
300	240	285	300	300	345	380	....	....
350	260	310	325	325	390	420	....	....
400	280	335	360	360	420	450	....	....
500	320	380	405	405	470	500	....	....

600	355	420	455	455	525	545	....	....
700	385	460	490	490	560	600	....	....
750	400	475	500	500	580	620	....	....
800	410	490	515	515	600	640	....	....
900	435	520	555	555	....	....	....	....
1000	455	545	585	585	680	730	....	....
1250	495	590	645	645	....	....	....	....
1500	520	625	700	700	785	....	....	....
1750	545	650	735	735	....	....	....	....
2000	560	665	775	775	840	....	....	....

TABLE 3 (310-14)—Allowable Ampacities of Insulated Aluminum and Copper-Clad Aluminum Conductors

This Table gives the allowable ampacities for not more than 3 conductors in a raceway or cable. When this number exceeds 3, the allowable ampacity of each conductor shall be reduced as follows:

4 to 6 conductors . . . . . 80 percent of value shown in Table
7 to 24 conductors . . . . . 70 percent of value shown in Table
25 to 42 conductors . . . . . 60 percent of value shown in Table
above 42 conductors . . . . . 50 percent of value shown in Table

Not More than Three Conductors in Raceway or Cable or Direct Burial (Based on Ambient Temperature of 30°C. 86°F.)

Size	Temperature Rating of Conductor. See Table 1 (310-2a)						
AWG MCM	60°C (140°F)	75°C (167°F)	85°C (185°F)	90°C (194°F)	110°C (230°F)	125°C (257°F)	200°C (392°F)
	TYPES RUW (12-2), T, TW	TYPES RH, RHW, RUH (12-2), THW, THWN XHHW	TYPES V, MI	TYPES TA, TBS, SA, AVB, SIS, RHH THHN XHHW**	TYPES AVA, AVL	TYPES AI (12-8), AIA	TYPES A (12-8), AA
12	15	15	25	25 †	25	30	30
10	25	25	30	30 †	35	40	45
8	30	40	40	40	45	50	55
6	40	50	55	55	60	65	75

**See footnotes p. 27.

Table 3 (310-14) (*Cont.*)

Size	Temperature Rating of Conductor. See Table 1 (310-2a)						
AWG MCM	60°C (140°F)	75°C (167°F)	85°C (185°F)	90°C (194°F)	110°C (230°F)	125°C (257°F)	200°C (392°F)
	TYPES RUW (12-2), T, TW	TYPES RH, RHW, RUH (12-2), THW, THWN XHHW	TYPES V, MI	TYPES TA, TBS, SA, AVB, SIS, RHH THHN XHHW**	TYPES AVA, AVL	TYPES AI (12-8), AIA	TYPES A (12-8), AA
4	55	65	70	70	80	90	95
3	65	75	80	80	95	100	115
*2	75	90	95	95	105	115	130
*1	85	100	110	110	125	135	150

*0	100	120	125	125	150	160	180
*00	115	135	145	145	170	180	200
*000	130	155	165	165	195	210	225
*0000	155	180	185	185	215	245	270
250	170	205	215	215	250	270	....
300	190	230	240	240	275	305	....
350	210	250	260	260	310	335	....
400	225	270	290	290	335	360	....
500	260	310	330	330	380	405	....
600	285	340	370	370	425	440	....
700	310	375	395	395	455	485	....
750	320	385	405	405	470	500	....
800	330	395	415	415	485	520	....
900	355	425	455	455	....	....	....
1000	375	445	480	480	560	600	....

*See footnotes p. 27.

Table 3 (310-14) (*Cont.*)

Size	Temperature Rating of Conductor. See Table 1 (310-2a)						
AWG MCM	60°C (140°F)	75°C (167°F)	85°C (185°F)	90°C (194°F)	110°C (230°F)	125°C (257°F)	200°C (392°F)
	TYPES RUW (12-2), T, TW	TYPES RH, RHW, RUH (12-2), THW, THWN XHHW	TYPES V, MI	TYPES TA, TBS, SA, AVB, SIS, RHH THHN XHHW**	TYPES AVA, AVL	TYPES AI (12-8), AIA	TYPES A (12-8), AA
1250	405	485	530	530	....	....	....
1500	435	520	580	580	650	....	....
1750	455	545	615	615	....	....	....
2000	470	560	650	650	705	....	....

These ampacities relate only to conductors described in Table 1 (310-2a).

*For three-wire, single-phase residential services, the allowable ampacity of RH, RHH, RHW, THW, and XHHW conductors shall be for sizes No. 2-100 Amp., No. 1-110 Amp., No. 1/0-125 Amp., No. 2/0-150 Amp., No. 3/0-175 Amp. and No. 4/0-200 Amp.

**For dry locations only. See Table 1 (310-2a).

†The ampacities for Type RHH, THHN, and XHHW conductors for sizes AWG 12 and 10 shall be the same as designated for 75°C conductors in this Table.

For ambient temperatures over 30°C, see Correction Factors, Table 6.

TABLE 4 (310-13)—Allowable Ampacities of Insulated Copper Conductors

Single Conductor in Free Air
(Based on Ambient Temperature of 30°C. 86°F.)

Size	Temperature Rating of Conductor. See Table 1 (310-2a).								
AWG MCM	60°C (140°F)	75°C (167°F)	85°C (185°F)	90°C (194°F)	110°C (230°F)	125°C (257°F)	200°C (392°F)	250°C (482°F)	
	TYPES RUW (14-2), T, TW	TYPES RH, RHW, RUH (14-2), THW, THWN XHHW	TYPES V, MI	TYPES TA, TBS, SA, AVB, SIS, FEP, FEPB, RHH, THHN, XHHW**	TYPES AVA, AVL	TYPES AI (14-8), AIA	TYPES A (14-8), AA, FEP* FEPB*	TYPE TFE (Nickel or nickel-coated copper only)	Bare and Covered Conductors

14	20	20	30	30†	40	40	45	60	30
12	25	25	40	40†	50	50	55	80	40
10	40	40	55	55†	65	70	75	110	55
8	55	65	70	70	85	90	100	115	70
6	80	95	100	100	120	125	135	210	100
4	105	125	135	135	160	170	180	285	130
3	120	145	155	155	180	195	210	335	150
2	140	170	180	180	210	225	240	390	175
1	165	195	210	210	245	265	280	450	205
0	195	230	245	245	285	305	325	545	235
00	225	265	285	285	330	355	370	605	275
000	260	310	330	330	385	410	430	725	320
0000	300	360	385	385	445	475	510	850	370
250	340	405	425	425	495	530	....	....	410
300	375	445	480	480	555	590	....	....	460
350	420	505	530	530	610	655	....	....	510
400	455	545	575	575	665	710	....	....	555
500	515	620	660	660	765	815	....	....	630

*See footnotes p. 31.

Table 4 (310-13) (*Cont.*)

Size	Temperature Rating of Conductor. See Table 1 (310-2a).								
AWG MCM	60°C (140°F)	75°C (167°F)	85°C (185°F)	90°C (194°F)	110°C (230°F)	125°C (257°F)	200°C (392°F)	250°C (482°F)	
	TYPES RUW (14-2), T, TW	TYPES RH, RHW, RUH (14-2), THW, THWN XHHW	TYPES V, MI	TYPES TA, TBS, SA, AVB, SIS, FEP, FEPB, RHH, THHN, XHHW**	TYPES AVA, AVL	TYPES AI (14-8), AIA	TYPES A (14-8), AA, FEP* FEPB*	TYPE TFE (Nickel or nickel-coated copper only)	Bare and Covered Conductors

600	575	690	740	740	855	910	....	....	710
700	630	755	815	815	940	1005	....	....	780
750	655	785	845	845	980	1045	....	....	810
800	680	815	880	880	1020	1085	....	....	845
900	730	870	940	940	....	....	....	....	905
1000	780	935	1000	1000	1165	1240	....	....	965
1250	890	1065	1130	1130	....	....	....	....	....
1500	980	1175	1260	1260	1450	....	....	....	1215
1750	1070	1280	1370	1370	....	....	....	....	....
2000	1155	1385	1470	1470	1715	....	....	....	1405

*Special use only. See Table 1 (310-2a).

**For dry locations only. See Table 1 (310-2a).

These ampacities relate only to conductors described in Table 1 (310-2a).

†The ampacities for Types FEP, FEPB, RHH, THHN, and XHHW conductors for sizes AWG 14, 12, and 10 shall be the same as designated for 75°C conductors in this Table.

For ambient temperatures over 30°C, see Correction Factors, Table 6.

TABLE 5 (310-15)—Allowable Ampacities of Insulated Aluminum and Copper-Clad Aluminum Conductors

Single Conductor in Free Air
(Based on Ambient Temperature of 30°C. 86°F.)

Size	Temperature Rating of Conductor. See Table 1 (310-2a).							
AWG MCM	60°C (140°F)	75°C (167°F)	85°C (185°F)	90°C (194°F)	110°C (230°F)	125°C (257°F)	200°C (392°F)	
	TYPES RUW (12-2), T, TW	TYPES RH, RHW, RUH (12-2), THW THWN XHHW	TYPES V, MI	TYPES TA, TBS, SA, AVB, SIS, RHH, THHN, XHHW*	TYPES AVA, AVL	TYPES AI (12-8), AIA	TYPES A (12-8), AA	Bare and Covered Conductors

12	20	20	30	30 †	40	40	45	30
10	30	30	45	45 †	50	55	60	45
8	45	55	55	55	65	70	80	55
6	60	75	80	80	95	100	105	80
4	80	100	105	105	125	135	140	100
3	95	115	120	120	140	150	165	115
2	110	135	140	140	165	175	185	135
1	130	155	165	165	190	205	220	160
0	150	180	190	190	220	240	255	185
00	175	210	220	220	255	275	290	215
000	200	240	255	255	300	320	335	250
0000	230	280	300	300	345	370	400	290
250	265	315	330	330	385	415	....	320
300	290	350	375	375	435	460	....	360
350	330	395	415	415	475	510	....	400
400	355	425	450	450	520	555	....	435
500	405	485	515	515	595	635	....	490

*See footnotes p. 35.

Table 5 (310-15) (*Cont.*)

Size	Temperature Rating of Conductor. See Table 1 (310-2a).							
AWG MCM	60°C (140°F)	75°C (167°F)	85°C (185°F)	90°C (194°F)	110°C (230°F)	125°C (257°F)	200°C (392°F)	
	TYPES RUW (12-2), T, TW	TYPES RH, RHW, RUH (12-2), THW THWN XHHW	TYPES V, MI	TYPES TA, TBS, SA, AVB, SIS, RHH, THHN, XHHW*	TYPES AVA, AVL	TYPES AI (12-8), AIA	TYPES A (12-8), AA	Bare and Covered Conductors

600	455	545	585	585	675	720	....	560
700	500	595	645	645	745	795	....	615
750	515	620	670	670	775	825	....	640
800	535	645	695	695	805	855	....	670
900	580	700	750	750	....	....	....	725
1000	625	750	800	800	930	990	....	770
1250	710	855	905	905	....	....	....	....
1500	795	950	1020	1020	1175	....	....	985
1750	875	1050	1125	1125	....	....	....	....
2000	960	1150	1220	1220	1425	....	....	1165

These ampacities relate only to conductors described in Table 1 (310-2a).

*For dry locations only. See Table 1 (310-2a).

†The ampacities for Types RHH, THHN, and XHHW conductors for sizes AWG 12 and 10 shall be the same as designated for 75°C conductors in this Table.

For ambient temperatures over 30°C, see Correction Factors, Table 6.

TABLE 6—Correction Factors for Ambient Temperatures above 30°C (86°F)

Where the room temperature is within 10°C of the maximum allowable operating temperature of the insulation, it is desirable to use an insulation with a higher maximum allowable operating temperature; although insulation can be used with a room temperature approaching the maximum allowable operating temperature limit if the allowable ampacity is reduced in accordance with the Correction Factors for different ambient temperatures as shown in the Table given below.

C.	F.	60°C (140°F)	75°C (167°F)	85°C (185°F)	90°C (194°F)	110°C (230°F)	125°C (257°F)	200°C (392°F)	250°C (482°F)
40	104	.82	.88	.90	.90	.94	.95	....	....
45	113	.71	.82	.85	.85	.90	.92	....	....
50	122	.58	.75	.80	.80	.87	.89	....	....
55	131	.41	.67	.74	.74	.83	.86	....	....

60	140	....	.58	.67	.67	.79	.83	.91	.95
70	158	....	.35	.52	.52	.71	.76	.87	.91
75	167	....	....	.43	.43	.66	.72	.86	.89
80	176	....	....	.30	.30	.61	.69	.84	.87
90	194	....	....	....	....	.50	.61	.80	.83
100	212	....	....	....	....	....	.51	.77	.80
120	248	....	....	....	....	....	....	.69	.72
140	284	....	....	....	....	....	....	.59	.59
160	320	....	....	....	....	....	....	....	.54
180	356	....	....	....	....	....	....	....	.50
200	392	....	....	....	....	....	....	....	.43
225	437	....	....	....	....	....	....	....	.30

TABLE 7 a and b (310-20c and 310-21)—Simplified Wiring Table and Typical Ambient Temperatures

The simplified wiring table, Table 7a (310-21), may be used for the selection of feeder and branch-circuit conductor sizes and insulation types when the following conditions and procedure are observed:

(1) This Table shall not be used to determine conductor ampacity. The ampere values listed apply to actual *diversified* continuous or non-continuous connected loads for which the demand factor is 80 percent or less.

(2) Determine load amperes, either continuous or non-continuous;
   (a) Continuous loads are expected to continue for three hours or longer;
   (b) Loads are non-continuous when 67 percent or less of the load is expected to be continuous.

(3) Select conductor sizes for load amperes—continuous and/or non-continuous, as the case may be—from Table 7a (310-21).

(4) Determine ambient temperature. Use 30°C (86°F) except where higher ambients may be expected, as tabulated in Table 7b (310-20(c)) following Table 7a.

(5) Select type of wire insulation from Table 1 (310-2a).

TABLE 7a (310-21)—Simplified Wiring Table
(Conductor Size*—6 or Fewer Conductors in Raceway or Cable)

Am-	Copper				Aluminum and Copper-Clad Aluminum			
	Non-Cont.		Continuous		Non-Cont.		Continuous	
peres	AWG	MCM	AWG	MCM	AWG	MCM	AWG	MCM
15	14		14		12		12	
20	12		12		10		10	
25	10		10		8		8	
30	10		10		8		8	
35	8		8		6		6	

*Neutral conductors may require special treatment.

Table 7a (310-21) (*Cont.*)

Am-	Copper				Aluminum and Copper-Clad Aluminum			
	Non-Cont.		Continuous		Non-Cont.		Continuous	
peres	AWG	MCM	AWG	MCM	AWG	MCM	AWG	MCM
40	8		8		6		6	
45	6		6		4		4	
50	6		6		4		4	
60	4		4		4		4	
70	4		4		3		3	
80	3		3		3		2	
90	3		2		2		1	
100	2		1		1		0	
110	1		0		0		2/0	
[illegible]	1		0		2/0		3/0	

150	0		2/0		3/0		4/0	
175	2/0		3/0		4/0			250
200	3/0		4/0			250		300
225	4/0			250		300		350
250		250		300		350		400
300		350		400		400		750
350		400		500		500		1000
400		500		750		750		
450		750		1000		1000		
500		750				1000		
600		1000						

TABLE 7b (310-20c)—Typical Ambient Temperatures

Location	Temperature	Minimum Rating of Required Conductor Insulation
Well ventilated, normally heated buildings	30°C (86°F)	* (See note below)
Buildings with such major heat sources as power stations or industrial processes	50°C (113°F)	75°C (167°F)
Poorly ventilated spaces such as attics	45°C (113°F)	75°C (167°F)

Furnaces and boiler rooms (min.)	40°C (104°F)	75°C (167°F)
(max.)	60°C (140°F)	90°C (194°F)
Outdoors in shade in air	40°C (104°F)	75°C (167°F)
In thermal insulation	45°C (113°F)	75°C (167°F)
Direct solar exposure	45°C (113°F)	75°C (167°F)
Places above 60°C (140°F)		110°C (230°F)

* Note: 60°C for up to and including No. 8 AWG copper and up to and including No. 6 AWG aluminum or copper-clad aluminum. 75°C for over No. 8 AWG copper and No. 6 AWG aluminum or copper-clad aluminum.

TABLE 8 (400-9b)—Ampacity of Flexible Cords

This gives the allowable ampacity for not more than 3 current-carrying conductors in a cord. If the number of current-carrying conductors in a cord is from 4 to 6, the allowable ampacity of each conductor shall be reduced to 80 percent of the values for not more than 3 current-carrying conductors in the Table. A conductor used for equipment grounding and a neutral conductor which carries only the unbalanced current from other conductors, as in the case of normally balanced circuits of 3 or more conductors, are not considered to be current-carrying conductors. Where a single conductor is used for both equipment grounding and to carry unbalanced current from other conductors, it shall not be considered to be a current-carrying conductor. (See Section 250-60.)

(Based on Ambient Temperature of 30°C (86°F). See Section 400-9 and Table 400-11)

Size AWG	Rubber Types TP, TS / Thermoplastic Types TPT, TST	Rubber Types PO, C, PD, E, EO, EN, S, SO, SRD, SJ, SJO, SV, SVO, SP / Thermoplastic Types ET, ETT, ETLB, ETP, ST, STO, SRDT, SJT, SJTO, SVT SVTO, SPT		Types AFS, AFSJ, HC, HPD, HSJ, HSJO, HS, HSO, HPN SVHT	Types AVPO AVPD	Cotton Types CFPD* / Asbestos Types AFC* AFPD*
		A†	B†			
27**	0.5	. .	. .	. .	. .	. .
18	.	7	10	10	17	6
17	. .		. .	12	. .	. .
16	. .	10	13	15	22	8

*See footnotes p. 47.

Table 8 (400-9b) (*Cont.*)

Size AWG	Rubber Types TP, TS / Thermo-plastic Types TPT, TST	Rubber Types PO, C, PD, E, EO, EN, S, SO, SRD, SJ, SJO, SV, SVO, SP / Thermo-plastic Types ET, ETT, ETLB, ETP, ST, STO, SRDT, SJT, SJTO, SVT SVTO, SPT		Types AFS, AFSJ, HC, HPD, HSJ, HSJO, HS, HSO, HPN SVHT	Types AVPO AVPD	Cotton Types CFPD* / Asbestos Types AFC* AFPD*
15	..		..	17	..	..
14	..	15	18	20	28	17
12	..	20	25	30	36	23
10	..	25	30	35	47	28

8	..	35	40	..	..	..
6	..	45	55	..	..	..
4	..	60	70	..	..	..
2	..	80	95	..	..	..

*These types are used almost exclusively in fixtures where they are exposed to high temperatures and ampere ratings are assigned accordingly.

**Tinsel Cord.

†The ampacities under subheading A are applicable to 3-conductor cords and other multi-conductor cords connected to utilization equipment so that only 3 conductors are current carrying. The ampacities under subheading B are applicable to 2-conductor cords and other multi-conductor cords connected to utilization equipment so that only 2 conductors are current carrying.

Note 1. Ultimate Insulation Temperature. In no case shall conductors be associated together in such way with respect to the kind of circuit, the wiring method employed, or the number of conductors that the limiting temperature of the conductors will be exceeded.

Note 2. SVHT made only in No. 18 and 17 AWG sizes.

TABLE 9
Percent of Cross-Section of Conduit and Tubing for Conductors

Number of Conductors	1	2	3	4	Over 4
All conductor types except lead-covered (new or rewiring)	53	31	40	40	40
Lead-covered conductors	55	30	40	38	35

Note 1. See Table 11, for number of conductors all of the same size in trade sizes of conduit ½ inch through 6 inch.

Note 2. For conductors larger than 750 MCM or for combinations of conductors of different sizes use Tables 4 through 8, Chapter 9 of the 1971 NEC, for dimensions of conductors, conduit and tubing.

Note 3. Where the calculated number of conductors, all of the same size, includes a decimal fraction, the next higher whole number shall be used where this decimal is 0.8 or larger.

Note 4. When bare conductors are permitted by other Sections of this Code, the dimensions for bare conductors in Table 16 may be used.

TABLE 10—Maximum Number of Fixture Wires in Trade Sizes of Conduit or Tubing (40 Percent Fill Based on Individual Diameters)

Conduit Trade Size (Inches)	½			¾			1			1¼			1½			2		
Wire Types	18	16	14	18	16	14	18	16	14	18	16	14	18	16	14	18	16	14
PTF, PTFF, PGFF, PGF, PFF, PF	23	18	14	40	31	24	65	50	39	115	90	70	157	122	95	257	200	156
TFFN, TFN	19	15		34	26		55	43		97	76		132	104		216	169	
SF-1	16			29			47			83			114			186		
SFF-1, FF-1, FFH-1	15			25			43			76			104			169		
CF	13	10	8	23	18	14	38	30	23	66	53	40	91	72	55	149	118	90
TF	11	10		20	18		32	30		57	53		79	72		129	118	
RFH-1, RF-1	11			20			32			57			79			129		
TFF	11	10		20	17		32	27		56	49		77	66		126	109	
AF	11	9	7	19	16	12	31	26	20	55	46	36	75	63	49	123	104	81
SFF-2	9	7	6	16	12	10	27	20	17	47	36	30	65	49	42	106	81	68
SF-2	9	8	6	16	14	11	27	23	18	47	40	32	65	55	43	106	90	71
FF-2, FFH-2	9	7		15	12		25	19		44	34		60	46		99	75	
RFH-2	7	5		12	10		20	16		36	28		49	38		80	62	
RF-2	7	6		12	10		20	16		36	29		49	40		80	65	

TABLE 11—Maximum Number of Conductors in Trade Sizes of Conduit or Tubing
(Tables 3A, 3B, 3C of Chapter 9, 1971 NEC)

The data in this table is based on Table 9.

Conduit Trade Size (Inches)		½	¾	1	1¼	1½	2	2½	3	3½	4	4½	5	6
Type Letters	Conductor Size AWG, MCM													
TW, T, RUH, RUW, XHHW (14 thru 8)	14	9	15	25	44	60	99	142						
	12	7	12	19	35	47	78	111	171					
	10	5	9	15	26	36	60	85	131	176				
	8	3	5	8	14	20	33	47	72	97	124			
RHW and RHH (without outer covering), THW	14	6	10	16	29	40	65	93	143	192				
	12	4	8	13	24	32	53	76	117	157				
	10	4	6	11	19	26	43	61	95	127	163			
	8	1	4	6	11	15	25	36	56	75	96	121	152	

TW,	6	1	2	4	7	10	16	23	36	48	62	78	97	141
T,	4	1	1	3	5	7	12	17	27	36	47	58	73	106
THW,	3	1	1	2	4	6	10	15	23	31	40	50	63	91
RUH (6 thru 2),	2	1	1	2	4	5	9	13	20	27	34	43	54	78
RUW (6 thru 2),	1		1	1	3	4	6	9	14	19	25	31	39	57
FEPB (6 thru 2),	0		1	1	2	3	5	8	12	16	21	27	33	49
RHW and	00		1	1	1	3	5	7	10	14	18	23	29	41
RHH (with-	000		1	1	1	2	4	6	9	12	15	19	24	35
out outer	0000			1	1	1	3	5	7	10	13	16	20	29
covering)	250			1	1	1	2	4	6	8	10	13	16	23
	300			1	1	1	2	3	5	7	9	11	14	20
	350				1	1	1	3	4	6	8	10	12	18
	400				1	1	1	2	4	5	7	9	11	16
	500				1	1	1	1	3	4	6	7	9	14
	600					1	1	1	3	4	5	6	7	11
	700					1	1	1	2	3	4	5	7	10
	750					1	1	1	2	3	4	5	6	9
	14	13	24	39	69	94	154							
	12	10	18	29	51	70	114	164						
THWN,	10	6	11	18	32	44	73	104	160					
	8	3	6	10	19	26	42	60	93	125	160			

TABLE 11 (*Cont.*)

Type Letters	Conductor Size AWG, MCM	½	¾	1	1¼	1½	2	2½	3	3½	4	4½	5	6
THHN, FEP (14 thru 2), FEPB (14 thru 8),	6	1	4	6	11	15	26	37	57	76	98	125	154	
	4	1	2	4	7	9	16	22	35	47	60	75	94	137
	3	1	1	3	6	8	13	19	29	39	51	64	80	116
	2	1	1	3	5	7	11	16	25	33	43	54	67	97
	1		1	1	3	5	8	12	18	25	32	40	50	72
XHHW (4 thru 500MCM)	0		1	1	3	4	7	10	15	21	27	33	42	61
	00		1	1	2	3	6	8	13	17	22	28	35	51
	000		1	1	1	3	5	7	11	14	18	23	29	42
	0000		1	1	1	2	4	6	9	12	15	19	24	35
	250			1	1	1	3	4	7	10	12	16	20	28
	300			1	1	1	3	4	6	8	11	13	17	24
	350			1	1	1	2	3	5	7	9	12	15	21
	400				1	1	1	3	5	6	8	10	13	19
	500				1	1	1	2	4	5	7	9	11	16
	600				1	1	1	1	3	4	5	7	9	13
	700					1	1	1	3	4	5	6	8	11
	750					1	1	1	2	3	4	6	7	11

XHHW	6	1	3	5	9	13	21	30	47	63	81	102	128	185
	600				1	1	1	1	3	4	5	7	9	13
	700					1	1	1	3	4	5	6	7	11
	750					1	1	1	2	3	4	6	7	10
	14	3	6	10	18	25	41	58	90	121	155			
	12	3	5	9	15	21	35	50	77	103	132			
RHW,	10	2	4	7	13	18	29	41	64	86	110	138		
	8	1	2	4	8	10	17	25	39	52	67	84	105	152
RHH	6	1	1	2	5	6	11	15	24	32	41	51	64	93
	4	1	1	1	3	5	8	12	18	24	31	39	50	72
(with	3	1	1	1	3	4	7	10	16	22	28	35	44	63
outer	2		1	1	3	4	6	9	14	19	24	31	38	56
covering)	1		1	1	1	3	5	7	11	14	18	23	29	42
	0		1	1	1	2	4	6	9	12	16	20	25	37
	00			1	1	1	3	5	8	11	14	18	22	32
	000			1	1	1	3	4	7	9	12	15	19	28
	0000			1	1	1	2	4	6	8	10	13	16	24
	250				1	1	1	3	5	6	8	11	13	19
	300				1	1	1	3	4	5	7	9	11	17
	350				1	1	1	2	4	5	6	8	10	15
	400				1	1	1	1	3	4	6	7	9	14
	500				1	1	1	1	3	4	5	6	8	11
	600					1	1	1	2	3	4	5	6	9
	700					1	1	1	1	3	3	4	6	8
	750						1	1	1	3	3	4	5	8

TABLE 12—Dimensions and Percent Area of Conduit and of Tubing (Area of conduit or tubing for the combinations of wires permitted in Table 9)

Trade Size	Internal Diameter Inches	Area—Square Inches								
			Not Lead Covered			Lead Covered				
		Total 100%	2 Cond. 31%	Over 2 Cond. 40%	1 Cond. 53%	1 Cond. 55%	2 Cond. 30%	3 Cond. 40%	4 Cond. 38%	Over 4 Cond. 35%
½	.622	.30	.09	.12	.16	.17	.09	.12	.11	.11
¾	.824	.53	.16	.21	.28	.29	.16	.21	.20	.19
1	1.049	.86	.27	.34	.46	.47	.26	.34	.33	.30
1¼	1.380	1.50	.47	.60	.80	.83	.45	.60	.57	.53
1½	1.610	2.04	.63	.82	1.08	1.12	.61	.82	.78	.71

2	2.067	3.36	1.04	1.34	1.78	1.85	1.01	1.34	1.28	1.18
2½	2.469	4.79	1.48	1.92	2.54	2.63	1.44	1.92	1.82	1.68
3	3.068	7.38	2.29	2.95	3.91	4.06	2.21	2.95	2.80	2.58
3½	3.548	9.90	3.07	3.96	5.25	5.44	2.97	3.96	3.76	3.47
4	4.026	12.72	3.94	5.09	6.74	7.00	3.82	5.09	4.83	4.45
4½	4.506	15.94	4.94	6.38	8.45	8.77	4.78	6.38	6.06	5.56
5	5.047	20.00	6.20	8.00	10.60	11.00	6.00	8.00	7.60	7.00
6	6.065	28.89	8.96	11.56	15.31	15.89	8.67	11.56	10.98	10.11

TABLE 13—Dimensions of Rubber-Covered and Thermoplastic-Covered Conductors

Size AWG MCM	Types RF-2, RFH-2, RH, RHH,*** RHW,*** SF-2		Types TF, T, THW,† TW, RUH,** RUW**		Types TFN, THHN, THWN		Types **** FEP, FEPB, TFE, PF, PGF, PTF		Type XHHW	
	Approx. Diam. Inches	Approx. Area Sq. In.	Approx. Diam. Inches	Approx. Area Sq. In.	Approx. Diam. Inches	Approx. Area Sq. In.	Approx. Diam. Inches	Approx. Area Sq. Inches	Approx. Diam. Inches	Approx. Area Sq. In.
Col. 1	Col. 2	Col. 3	Col. 4	Col. 5	Col. 6	Col. 7	Col. 8	Col. 9	Col. 10	Col. 11
18	.146	.0167	.106	.0088	.089	.0064	.081	.0052	...	....
16	.158	.0196	.118	.0109	.100	.0079	.092	.0066	...	....
14	30 mils .171	.0230	.131	.0135	.105	.0087	.105 .105	.0087 .0087		
14	45 mils .204*	.0327*	...	....	...	....	... ...	....		
14	....	....	.162†	.0206†	...	....	... ...	....	.129	.0131
12	30 mils .188	.0278	.148	.0172	.122	0117	.121 .121	.0115 .0115		
12	45 mils .221*	.0384*	...	....	...	....	... ...	....		
12	....	....	.179†	.0251†	...	....	... ...	....	.146	.0167
10	...... .242	.0460	.168	.0224	.153	.0184	.142 .142	.0159 .0159		
10	....	....	.199†	.0311†	...	....	... ...	....	.166	.0216
8	...... .311	.0760	.228	.0408	.201	.0317	.189 .169	.0280 .0225		
8	....	....	.259†	.0526†	...	....	... ...	....	.224	.0394

6	.397	.1238	.323	.0819	.257	.0519	.244	.302	.0467	.0716	.282	.0625
4	.452	.1605	.372	.1087	.328	.0845	.292	.350	.0669	.0962	.328	.0845
3	.481	.1817	.401	.1263	.356	.0995	.320	.378	.0803	.1122	.356	.0995
2	.513	.2067	.433	.1473	.388	.1182	.352	.410	.0973	.1316	.388	.1182
1	.588	.2715	.508	.2027	.450	.1590	.420	...	.1385	....	.450	.1590
0	.629	.3107	.549	.2367	.491	.1893	.462	...	.1676	....	.491	.1893
00	.675	.3578	.595	.2781	.537	.2265	.498	...	.1974	....	.537	.2265
000	.727	.4151	.647	.3288	.588	.2715	.560	...	.2463	....	.588	.2715
0000	.785	.4840	.705	.3904	.646	.3278	.618	...	.2999	....	.646	.3278
250	.868	.5917	.788	.4877	.716	.4026	...		....		.716	.4026
300	.933	.6837	.843	.5581	.771	.4669	...		....		.771	.4669
350	.985	.7620	.895	.6291	.822	.5307	...		....		.822	.5307
400	1.032	.8365	.942	.6969	.869	.5931	...		....		.869	.5931
500	1.119	.9834	1.029	.8316	.955	.7163	...		....		.955	.7163
600	1.233	1.1940	1.143	1.0261	1.058	.8792	...		....		1.073	.9043
700	1.304	1.3355	1.214	1.1575	1.129	1.0011	...		....		1.145	1.0297
750	1.339	1.4082	1.249	1.2252	1.163	1.0623	...		....		1.180	1.0936
800	1.372	1.4784	1.282	1.2908	1.196	1.1234	...		....		1.210	1.1499
900	1.435	1.6173	1.345	1.4208	1.259	1.2449	...		....		1.270	1.2668

*See footnotes p. 58.

TABLE 13 (*Cont.*)

Size AWG MCM	Types RF-2, RFH-2, RH, RHH,*** RHW,*** SF-2		Types TF, T, THW,† TW, RUH,** RUW**		Types TFN, THHN, THWN		Types **** FEP, FEPB, TFE, PF, PGF, PTF		Type XHHW	
	Approx. Diam. Inches	Approx. Area Sq. In.	Approx. Diam. Inches	Approx. Area Sq. In.	Approx. Diam. Inches	Approx. Area Sq. In.	Approx. Diam. Inches	Approx. Area Sq. Inches	Approx. Diam. Inches	Approx. Area Sq. In.
Col. 1	Col. 2	Col. 3	Col. 4	Col. 5	Col. 6	Col. 7	Col. 8	Col. 9	Col. 10	Col. 11
1000	1.494	1.7531	1.404	1.5482	1.317	1.3623	...	....	1.330	1.3893
1250	1.676	2.2062	1.577	1.9532	...	....	...	....	1.500	1.7672
1500	1.801	2.5475	1.702	2.2748	...	....	...	....	1.620	2.0612
1750	1.916	2.8895	1.817	2.5930	...	....	...	....	1.740	2.3779
2000	2.021	3.2079	1.922	2.9013	...	....	...	....	1.840	2.6590

*The dimensions of Types RHH and RHW.

**No. 14 to No. 2.

†Dimensions of THW in sizes 14 to 8. No. 6 THW and larger is the same dimension as T.

***Dimensions of RHH and RHW without outer covering are the same as THW.

No. 18 to No. 8, solid; No. 6 and larger, stranded.

****In Columns 8 and 9 the values shown for sizes No. 1 thru 0000 are for TFE only. The right-hand values in Columns 8 and 9 are for FEPB only.

TABLE 14—Dimensions of Lead-Covered Conductors

(Types RL, RHL, RUL)

Size AWG-MCM	Single Conductor		Two Conductor		Three Conductor	
	Diam. Inches	Area Sq. Ins.	Diam. Inches	Area Sq. Ins.	Diam. Inches	Area Sq. Ins.
14	.28	.062	.28 x .47	.115	.59	.273
12	.29	.066	.31 x .54	.146	.62	.301
10	.35	.096	.35 x .59	.180	.68	.363
8	.41	.132	.41 x .71	.255	.82	.528
6	.49	.188	.49 x .86	.369	.97	.738
4	.55	.237	.54 x .96	.457	1.08	.916
2	.60	.283	.61 x 1.08	.578	1.21	1.146
1	.67	.352	.70 x 1.23	.756	1.38	1.49

Table 14 *(Cont.)*

Size AWG-MCM	Single Conductor		Two Conductor		Three Conductor	
	Diam. Inches	Area Sq. Ins.	Diam. Inches	Area Sq. Ins.	Diam. Inches	Area Sq. Ins.
0	.71	.396	.74 x 1.32	.859	1.47	1.70
00	.76	.454	.79 x 1.41	.980	1.57	1.94
000	.81	.515	.84 x 1.52	1.123	1.69	2.24
0000	.87	.593	.90 x 1.64	1.302	1.85	2.68
250	.98	.754	....	.....	2.02	3.20
300	1.04	.85	....	.....	2.15	3.62
350	1.10	.95	....	.....	2.26	4.02
400	1.14	1.02	....	.....	2.40	4.52
500	1.23	1.18	....	.....	2.59	5.28

The above cables are limited to straight runs or with nominal offsets equivalent to not more than two quarter bends.

Note: No. 14 to No. 8, solid conductors: No. 6 and larger, stranded conductors. Data for 30-mil insulation not yet compiled.

TABLE 15—Dimensions of Asbestos-Varnished-Cambric Insulated Conductors

(Types AVA, AVB, AVL)

Size AWG MCM	Type AVA		Type AVB		Type AVL	
	Approx. Diam. Inches	Approx. Area Sq. Ins.	Approx. Diam. Inches	Approx. Area Sq. Ins.	Approx. Diam. Inches	Approx. Area Sq. Ins.
14	.245	.047	.205	.033	.320	.080
12	.265	.055	.225	.040	.340	.091
10	.285	.064	.245	.047	.360	.102
8	.310	.075	.270	.057	.390	.119
6	.395	.122	.345	.094	.430	.145
4	.445	.155	.395	.123	.480	.181
2	.505	.200	.460	.166	.570	.255
1	.585	.268	.540	.229	.620	.300

0	.625	.307	.580	.264	.660	.341
00	.670	.353	.625	.307	.705	.390
000	.720	.406	.675	.358	.755	.447
0000	.780	.478	.735	.425	.815	.521
250	.885	.616	.855	.572	.955	.715
300	.940	.692	.910	.649	1.010	.800
350	.995	.778	.965	.731	1.060	.885
400	1.040	.850	1.010	.800	1.105	.960
500	1.125	.995	1.095	.945	1.190	1.118
550	1.165	1.065	1.135	1.01	1.265	1.26
600	1.205	1.140	1.175	1.09	1.305	1.34
650	1.240	1.21	1.210	1.15	1.340	1.41
700	1.275	1.28	1.245	1.22	1.375	1.49
750	1.310	1.35	1.280	1.29	1.410	1.57
800	1.345	1.42	1.315	1.36	1.440	1.63

Table 15 *(Cont.)*

Size AWG MCM	Type AVA		Type AVB		Type AVL	
	Approx. Diam. Inches	Approx. Area Sq. Ins.	Approx. Diam. Inches	Approx. Area Sq. Ins.	Approx. Diam. Inches	Approx. Area Sq. Ins.
850	1.375	1.49	1.345	1.43	1.470	1.70
900	1.405	1.55	1.375	1.49	1.505	1.78
950	1.435	1.62	1.405	1.55	1.535	1.85
1000	1.465	1.69	1.435	1.62	1.565	1.93

Note: No. 14 to No. 8, solid, No. 6 and larger, stranded; except AVL where all sizes are stranded.

TABLE 16—Properties of Conductors

Size AWG MCM	Area Cir. Mils	Concentric Lay Stranded Conductors		Bare Conductors		D. C. Resistance Ohms/M Ft. At 25°C. 77°F.		
						Copper		Aluminum
		No. Wires	Diam. Each Wire Inches	Diam. Inches	*Area Sq. Inches	Bare Cond.	Tin'd. Cond.	
18	1620	Solid	.0403	.0403	.0013	6.51	6.79	10.7
16	2580	Solid	.0508	.0508	.0020	4.10	4.26	6.72
14	4110	Solid	.0641	.0641	.0032	2.57	2.68	4.22
12	6530	Solid	.0808	.0808	.0051	1.62	1.68	2.66
10	10380	Solid	.1019	.1019	.0081	1.018	1.06	1.67
8	16510	Solid	.1285	.1285	.0130	.6404	.659	1.05

*See footnote p. 69.

Table 16 *(Cont.)*

Size AWG MCM	Area Cir. Mils	Concentric Lay Stranded Conductors		Bare Conductors		D. C. Resistance Ohms/M Ft. At 25°C. 77°F.		
						Copper		Alumni-num
		No. Wires	Diam. Each Wire Inches	Diam. Inches	*Area Sq. Inches	Bare Cond.	Tin'd. Cond.	
6	26240	7	.0612	.184	.027	.410	.427	.674
4	41740	7	.0772	.232	.042	.259	.269	.424
3	52620	7	.0867	.260	.053	.205	.213	.336
2	66360	7	.0974	.292	.067	.162	.169	.266
1	83690	19	.0664	.332	.087	.129	.134	.211

0	105600	19	.0745	.372	.109	.102	.106	.168
00	133100	19	.0837	.418	.137	.0811	.0843	.133
000	167800	19	.0940	.470	.173	.0642	.0668	.105
0000	211600	19	.1055	.528	.219	.0509	.0525	.0836
250	250000	37	.0822	.575	.260	.0431	.0449	.0708
300	300000	37	.0900	.630	.312	.0360	.0374	.0590
350	350000	37	.0973	.681	.364	.0308	.0320	.0505
400	400000	37	.1040	.728	.416	.0270	.0278	.0442
500	500000	37	.1162	.813	.519	.0216	.0222	.0354
600	600000	61	.0992	.893	.626	.0180	.0187	.0295
700	700000	61	.1071	.964	.730	.0154	.0159	.0253
750	750000	61	.1109	.998	.782	.0144	.0148	.0236
800	800000	61	.1145	1.030	.833	.0135	.0139	.0221
900	900000	61	.1215	1.090	.933	.0120	.0123	.0197

*See footnote p. 69.

Table 16 *(Cont.)*

Size AWG MCM	Area Cir. Mils	Concentric Lay Stranded Conductors		Bare Conductors		D. C. Resistance Ohms/M Ft. At 25°C. 77°F.		
						Copper		Alumninum
		No. Wires	Diam. Each Wire Inches	Diam. Inches	*Area Sq. Inches	Bare Cond.	Tin'd. Cond.	
1000	1000000	61	.1280	1.150	1.039	.0108	.0111	.0177
1250	1250000	91	.1172	1.289	1.305	.00863	.00888	.0142
1500	1500000	91	.1284	1.410	1.561	.00719	.00740	.0118
1750	1750000	127	.1174	1.526	1.829	.00616	.00634	.0101
2000	2000000	127	.1255	1.630	2.087	.00539	.00555	.00885

*Area given is that of a circle having a diameter equal to the over-all diameter of a stranded conductor.

The values given in the Table are those given in Handbook 100 of the National Bureau of Standards except that those shown in the 8th column are those given in Specification B33 of the American Society for Testing and Materials, and those shown in the 9th column are those given in Standard No. S-19-81 of the Insulated Power Cable Engineers Association and Standard No. WC3-1964 of the National Electrical Manufacturers Association.

The resistance values given in the last three columns are applicable only to direct current. When conductors larger than No. 4/0 are used with alternating current the multiplying factors in Table 9, Chapter 9 of 1971 NEC should be used to compensate for skin effect.

TABLE 17—Multiplying Factors for Converting DC Resistance to 60-Hertz AC Resistance

Size	Multiplying Factor			
	For Nonmetallic Sheathed Cables in Air or Nonmetallic Conduit		For Metallic Sheathed Cables or all Cables in Metallic Raceways	
	Copper	Aluminum	Copper	Aluminum
Up to 3 AWG	1.	1.	1.	1.
2	1.	1.	1.01	1.00
1	1.	1.	1.01	1.00
0	1.001	1.000	1.02	1.00
00	1.001	1.001	1.03	1.00
000	1.002	1.001	1.04	1.01
0000	1.004	1.002	1.05	1.01

250 MCM	1.005	1.002	1.06	1.02
300 MCM	1.006	1.003	1.07	1.02
350 MCM	1.009	1.004	1.08	1.03
400 MCM	1.011	1.005	1.10	1.04
500 MCM	1.018	1.007	1.13	1.06
600 MCM	1.025	1.010	1.16	1.08
700 MCM	1.034	1.013	1.19	1.11
750 MCM	1.039	1.015	1.21	1.12
800 MCM	1.044	1.017	1.22	1.14
1000 MCM	1.067	1.026	1.30	1.19
1250 MCM	1.102	1.040	1.41	1.27
1500 MCM	1.142	1.058	1.53	1.36
1750 MCM	1.185	1.079	1.67	1.46
2000 MCM	1.233	1.100	1.82	1.56

TABLE 18 (430-147)
Full-Load Currents for Direct-Current Motors

The following values of full-load currents are for motors running at base speed.

HP	120V	240V
1/4	2.9	1.5
1/3	3.6	1.8
1/2	5.2	2.6
3/4	7.4	3.7
1	9.4	4.7
1 1/2	13.2	6.6
2	17	8.5
3	25	12.2
5	40	20
7 1/2	58	29
10	76	38
15		55
20		72
25		89
30		106
40		140
50		173
60		206
75		255
100		341
125		425
150		506
200		675

TABLE 19 (430-148)
Full-Load Currents for Single-Phase Alternating-Current Motors

The following values of full-load currents are for motors running at usual speeds and motors with normal torque characteristics. Motors built for especially low speeds or high torques may have higher full-load currents, and multispeed motors will have full-load current varying with speed, in which case the nameplate current ratings shall be used.

To obtain full-load currents of 208- and 200-volt motors, increase corresponding 230-volt motor full-load currents by 10 and 15 percent, respectively.

The voltages listed are rated motor voltages. Corresponding nominal system voltages are 110 to 120 and 220 to 240.

HP	115V	230V
1/6	4.4	2.2
1/4	5.8	2.9
1/3	7.2	3.6
1/2	9.8	4.9
3/4	13.8	6.9
1	16	8
1 1/2	20	10
2	24	12
3	34	17
5	56	28
7 1/2	80	40
10	100	50

TABLE 20 (430-149)—Full-Load Currents for Two-Phase Alternating-Current Motors (4-wire)

The following values of full-load current are for motors running at speeds usual for belted motors and motors with normal torque characteristics. Motors built for especially low speeds or high torques may require more running current, and multispeed motors will have full-load current varying with speed, in which case the nameplate current rating shall be used. Current in common conductor of 2-phase, 3-wire system will be 1.41 times value given.

The voltages listed are rated motor voltages. Corresponding nominal system voltages are 110 to 120, 220 to 240, 440 to 480 and 550 to 600 volts.

HP	Induction Type Squirrel-Cage and Wound Rotor Amperes					Synchronous Type †Unity Power Factor Amperes			
	115V	230V	460V	575V	2300V	220V	440V	550V	2300V
½	4	2	1	.8					
¾	4.8	2.4	1.2	1.0					
1	6.4	3.2	1.6	1.3					
1½	9	4.5	2.3	1.8					
2	11.8	5.9	3	2.4					
3		8.3	4.2	3.3					
5		13.2	6.6	5.3					
7½		19	9	8					
10		24	12	10					

†See p. 77.

Table 20 *(Cont.)*

	Induction Type Squirrel-Cage and Wound Rotor Amperes					Synchronous Type †Unity Power Factor Amperes			
HP	115V	230V	460V	575V	2300V	220V	440V	550V	2300V
15		36	18	14					
20		47	23	19					
25		59	29	24		47	24	19	
30		69	35	28		56	29	23	
40		90	45	36		75	37	31	
50		113	56	45		94	47	38	
60		133	67	53	14	111	56	44	11
75		166	83	66	18	140	70	57	13
100		218	109	87	23	182	93	74	17

125		270	135	108	28	228	114	93	22
150		312	156	125	32		137	110	26
200		416	208	167	43		182	145	35

†For 90 and 80 percent power factor the above figures should be multiplied by 1.1 and 1.25 respectively.

TABLE 21 (430-150)—Full-Load Currents* for Three-Phase Alternating-Current Motors

	Induction Type Squirrel-Cage and Wound Rotor Amperes					Synchronous Type †Unity Power Factor Amperes			
HP	115V	230V	460V	575V	2300V	220V	440V	550V	2300V
½	4	2	1	.8					
¾	5.6	2.8	1.4	1.1					
1	7.2	3.6	1.8	1.4					
1½	10.4	5.2	2.6	2.1					
2	13.6	6.8	3.4	2.7					
3		9.6	4.8	3.9					

5		15.2	7.6	6.1					
7½		22	11	9					
10		28	14	11					
15		42	21	17					
20		54	27	22					
25		58	34	27		54	27	22	
30		30	40	32		65	33	26	
40		104	52	41		86	43	35	
50		130	65	52		108	54	44	
60		154	77	62	16	128	64	51	12
75		192	96	77	20	161	81	65	15
100		248	124	99	26	211	106	85	20

*See footnotes p. 81.

Table 21 *(Cont.)*

	Induction Type Squirrel-Cage and Wound Rotor Amperes					Synchronous Type †Unity Power Factor Amperes			
HP	**115V**	**230V**	**460V**	**575V**	**2300V**	**220V**	**440V**	**550V**	**2300V**
125		312	156	125	31	264	132	106	25
150		360	180	144	37		158	127	30
200		480	240	192	49		210	168	40

For full-load currents of 208- and 200-volt motors increase the corresponding 230-volt motor full-load current by 10 and 15 percent respectively.

*These values of full-load current are for motors running at speeds usual for belted motors and motors with normal torque characteristics. Motors built for especially low speeds or high torques may require more running current, and multispeed motors will have full-load current varying with speed, in which case the nameplate current rating shall be used.

†For 90 and 80 percent power factor the above figures shall be multiplied by 1.1 and 1.25 respectively.

The voltages listed are rated motor voltages. Corresponding nominal system voltages are 110 to 120, 220 to 240, 440 to 480, and 550 to 600 volts.

TABLE 22 (430-7b)
Locked-Rotor Indicating Code Letters

Code Letter	Kilovolt-Amperes per Horsepower with Locked Rotor
A	0 — 3.14
B	3.15 — 3.54
C	3.55 — 3.99
D	4.0 — 4.49
E	4.5 — 4.99
F	5.0 — 5.59
G	5.6 — 6.29
H	6.3 — 7.09
J	7.1 — 7.99
K	8.0 — 8.99
L	9.0 — 9.99
M	10.0 — 11.19
N	11.2 — 12.49
P	12.5 — 13.99
R	14.0 — 15.99
S	16.0 — 17.99
T	18.0 — 19.99
U	20.0 — 22.39
V	22.4 — and up

Explanation: The starting current of a motor is usually much larger than its normal running current. In order to select over-current protection (for alternating current motors), motors larger than 1/2 hp are identified on their nameplates by a code letter indicating the locked-rotor or starting current of the motor. Table 22 is an adopted standard and shows the motor kva input for locked-rotor or starting conditions according to specified code letters.

Thus, a 10 hp, 220-volt, 3-phase induction motor with identifying Code Letter D would take from 40 to 44.9 kva from the lines when starting at rated voltage. This corresponds to a minimum line current of $\frac{40 \times 1000}{\sqrt{3} \times 220} = 105$ amperes; or about 3.9 times rated current for this motor as given in Table 21.

TABLE 23 (430-151)—Locked-Rotor Current Conversion Table

As Determined from Horsepower and Voltage Rating
For Use Only With Sections 430-110, 440-12, and 440-41.

Conversion Table

Max HP Rating	Motor Locked-Rotor Current Amperes					
	Single Phase		Two or Three Phase			
	115 V	230 V	115 V	230 V	460 V	575 V
½	58.8	29.4	24	12	6	4.8
¾	82.8	41.4	33.6	16.8	8.4	6.6
1	96	48	42	21	10.8	8.4
1½	120	60	60	30	15	12
2	144	72	78	39	19.8	15.6
3	204	102	—	54	27	24
5	336	168	—	90	45	36

7½	480	240	—	132	66	54
10	600	300	—	162	84	66
15	—	—	—	240	120	96
20	—	—	—	312	156	126
25	—	—	—	384	192	156
30	—	—	—	468	234	186
40	—	—	—	624	312	246
50	—	—	—	750	378	300
60	—	—	—	900	450	360
75	—	—	—	1110	558	444
100	—	—	—	1476	738	588
125	—	—	—	1860	930	744
150	—	—	—	2160	1080	864
200	—	—	—	2880	1440	1152

TABLE 24 (430-152)—Maximum Rating or Setting of Motor Branch-Circuit Protective Devices

Type of Motor	Percent of Full-Load Current: Nontime Delay Fuse	Percent of Full-Load Current: Dual-Element (Time-Delay) Fuse	Percent of Full-Load Current: Instantaneous Type Breaker	Percent of Full-Load Current: Time-Limit Breaker
Single-phase, all types				
No code letter	300	175	700	250
All AC single-phase and polyphase squirrel-cage and synchronous motors with full-voltage, resistor or reactor starting:				
No code letter	300	175	700	250
Code letter F to V	300	175	700	250

Code letter B to E	250	175	700	200
Code letter A	150	150	700	150
All AC squirrel-cage and synchronous motors with autotransformer starting:				
Not more than 30 amps				
No code letter	250	175	700	200
More than 30 amps				
No code letter	200	175	700	200
Code letter F to V	250	175	700	200
Code letter B to E	200	175	700	200
Code letter A	150	150	700	150
High-reactance squirrel-cage				
Not more than 30 amps				
No code letter	250	175	700	250
More than 30 amps				
No code letter	200	175	700	200

Table 24 *(Cont.)*

Type of Motor	Percent of Full-Load Current			
	Nontime Delay Fuse	Dual-Element (Time-Delay) Fuse	Instantaneous Type Breaker	Time-Limit Breaker
Wound-rotor — No code letter	150	150	700	150
Direct-current				
No more than 50 hp				
No code letter	150	150	250	150
More than 50 hp				
No code letter	150	150	175	150

For explanation of Code Letter Marking, see Table 22 (430-7b).

For certain exceptions to the values specified see Sections 430-52, -54. The values given in the last column also cover the ratings of nonadjustable time-limit types of circuit breakers which may be modified as in Section 430-52.

Synchronous motors of the low-torque, low-speed type (usually 450 RPM or lower), such as are used to drive reciprocating compressors, pumps, etc., which start unloaded, do not require a fuse rating or circuit-breaker setting in excess of 200 percent of full-load current.

TABLE 25 (220-5)—Demand Loads for Household Electric Ranges, Wall-Mounted Ovens, Counter-Mounted Cooking Units and Other Household Cooking Appliances over 1¾ kW Rating Column A to be used in all cases except as otherwise permitted in Note 4 below.

NUMBER OF APPLIANCES	Maximum Demand (See Notes)	Demand Factors (See Note 4)	
	COLUMN A (Not over 12 kW Rating)	COLUMN B (Less than 3½ kW Rating)	COLUMN C (3½ kw to 8¾ kW Rating)
1	8 kW	80%	80%
2	11 kW	75%	65%
3	14 kW	70%	55%
4	17 kW	66%	50%
5	20 kW	62%	45%
6	21 kW	59%	43%
7	22 kW	56%	40%

8	23 kW	53%	36%
9	24 kW	51%	35%
10	25 kW	49%	34%
11	26 kW	47%	32%
12	27 kW	45%	32%
13	28 kW	43%	32%
14	29 kW	41%	32%
15	30 kW	40%	32%
16	31 kW	39%	28%
17	32 kW	38%	28%
18	33 kW	37%	28%
19	34 kW	36%	28%
20	35 kW	35%	28%
21	36 kW	34%	26%
22	37 kW	33%	26%
23	38 kW	32%	26%
24	39 kW	31%	26%
25	40 kW	30%	26%
26-30	15 kW plus 1 kW	30%	24%
31-40	for each range	30%	22%

Table 25 *(Cont.)*

NUMBER OF APPLIANCES	Maximum Demand (See Notes)	Demand Factors (See Note 4)	
	COLUMN A (Not over 12 kW Rating)	COLUMN B (Less than 3½ kW Rating)	COLUMN C (3½ kw to 8¾ kW Rating)
41-50	25 kW plus ¾ kW for each range	30%	20%
51-60	25 kW plus ¾ kW for each range	30%	18%
61 & over	25 kW plus ¾ kW for each range	30%	16%

Note 1. Over 12 kW to 27 kW ranges all of same rating. For ranges, individually rated more than 12 kW but not more than 27 kW, the maximum demand in Column A shall be increased 5 percent for each additional kW of rating or major fraction thereof by which the rating of individual ranges exceeds 12 kW.

Note 2. Over 12 kW to 27 kW ranges *of unequal ratings.* For ranges individually rated more than 12 kW and of different ratings but none exceeding 27 kW an average value of rating shall be calculated by adding together the ratings of all ranges to obtain the total connected load (using 12 kW for any range rated less than 12 kW) and dividing by the total number of ranges; and then the maximum demand in Column A shall be increased 5 percent for each kW or major fraction thereof by which this average value exceeds 12 kW.

Note 3. This Table does not apply to commercial ranges. See Table 220-6(a) of 1971 NEC for demand factors for commercial cooking equipment.

Note 4. Over 1¾ kW to 8¾ kW. In lieu of the method provided in Column A, loads rated more than 1¾ kW but not more than 8¾ kW may be considered as the sum of the nameplate ratings of all the loads, multiplied by the demand factors specified in Columns B or C for the given number of loads.

Note 5. Branch-Circuit Load. Branch-circuit load for one range may be computed in accordance with Table 25(220-5). The branch-circuit load for one wall-mounted oven or one counter-mounted cooking unit shall be the nameplate rating of the appliance. The branch-circuit load for a counter-mounted cooking unit and not more than two wall-mounted ovens, all supplied from a single branch circuit and located in the same room shall be computed by adding the nameplate ratings of the individual appliances and treating this total as equivalent to one range.

TABLE 26—Voltage Drop Data*

Good wiring demands that the voltage drop in feeders as well as in branch circuit conductors be kept at reasonably low limits (approximately 2 percent from meter to appliance). Table 26 gives data for rapid calculations for copper conductors in steel conduit.

Rule for finding voltage drop:—Multiply the current (amperes) by the distance (feet in one conductor) by the figure in Table 26 for the size wire used and the type of system. Point off 6 places in this figure to give the number of volts drop.

Illustration:—For a No. 8 wire carrying 40 amperes in a 3-phase system at 90 percent power factor, the number from Table 26 is 1189. For a one-way run of 100 feet, the product is: 40 × 100 × 1189 = 4,756,000 or 4.756 volts drop.

Wire Size	Power Factor %	A-c Single Phase	A-c Three Phase	D-c	Wire Size	Power Factor %	A-c Single Phase	A-c Three Phase	D-c
14 Awg	100	5880	5090	5880	250 MCM	100	102	89	97
	90	5360	4640			90	136	117	
	80	4790	4150			80	143	123	
	70	4230	3660			70	143	124	
	60	3650	3160			60	141	122	

12 Awg	100	3690	3190	3690	300 MCM	100	86	75	81
	90	3380	2930			90	121	104	
	80	3030	2620			80	128	111	
	70	2680	2320			70	131	113	
	60	2320	2010			60	130	113	
10 Awg	100	2320	2010	2320	350 MCM	100	74	64	69
	90	2150	1861			90	109	95	
	80	1935	1675			80	118	102	
	70	1718	1487			70	122	105	
	60	1497	1296			60	122	106	
8 Awg	100	1462	1255	1462	400 MCM	100	66	57	60
	90	1373	1139			90	101	88	
	80	1248	1081			80	111	96	
	70	1117	969			70	115	99	
	60	981	849			60	116	101	
6 Awg	100	918	795	918	500 MCM	100	54	47	48
	90	882	764			90	89	78	
	80	812	703			80	99	86	
	70	734	636			70	105	91	
	60	653	565			60	108	93	

*Courtesy American Iron and Steel Institute.

Table 26 *(Cont.)*

Wire Size	Power Factor %	A-c Single Phase	A-c Three Phase	D-c	Wire Size	Power Factor %	A-c Single Phase	A-c Three Phase	D-c
4 Awg	100	578	501	578	600 MCM	100	47	41	40
	90	571	494			90	83	72	
	80	533	462			80	93	81	
	70	489	423			70	99	86	
	60	440	381			60	103	89	
2 Awg	100	367	318	363	700 MCM	100	41	36	34
	90	379	328			90	77	67	
	80	361	313			80	88	76	
	70	337	292			70	94	82	
	60	309	268			60	98	85	
1 Awg	100	291	252	288	750 MCM	100	39	34	32
	90	311	269			90	75	65	
	80	299	259			80	86	75	
	70	284	246			70	93	81	
	60	264	229			60	97	84	
0 Awg	100	233	202	229	800 MCM	100	37	32	30
	90	257	222			90	73	63	
	80	252	218			80	84	73	
	70	241	209			70	91	79	
	60	227	196			60	95	82	

00 Awg	100	187	162	181	900 MCM	100	34	29	27
	90	213	184			90	69	61	
	80	212	183			80	81	71	
	70	206	178			70	88	77	
	60	196	169			60	93	81	
000 Awg	100	149	129	144	1000 MCM	100	31	27	24
	90	179	155			90	67	58	
	80	181	156			80	79	68	
	70	177	153			70	86	75	
	60	171	148			60	91	78	
0000 Awg	100	121	104	114					
	90	152	131						
	80	156	135						
	70	155	134						
	60	151	131						

TABLE 27

American Wire Gauge

The American Wire Gauge is standard for copper wire and applies to all cases where size of copper or aluminum wire is mentioned in any wire gauge number. The complete wire table is based on a geometrical formula such that for all practical purposes, the area doubles for every third size larger. For every ten wire sizes larger, the area increases by ten times. It will be noted that No. 10 AWG has an area of approximately 10,000 circular mils. (Table 16 lists only even-numbered sizes since these are most commonly used.)

For some purposes, the following equivalence of wire sizes will be helpful.

0000 =	2-0 =	4-3 =	8-6 =	16-9 =	32-12 =	64-15
000 =	2-1 =	4-4 =	8-7 =	16-10 =	32-13 =	64-16
00 =	2-2 =	4-5 =	8-8 =	16-11 =	32-14 =	64-17
0 =	2-3 =	4-6 =	8-9 =	16-12 =	32-15	
1 =	2-4 =	4-7 =	8-10 =	16-13 =	32-16	
2 =	2-5 =	4-8 =	8-11 =	16-14 =	32-17	
3 =	2-6 =	4-9 =	8-12 =	16-15 =	32-18	
4 =	2-7 =	4-10 =	8-13 =	16-16		
5 =	2-8 =	4-11 =	8-14 =	16-17		
6 =	2-9 =	4-12 =	8-15 =	16-18		
7 =	2-10 =	4-13 =	8-16			
8 =	2-11 =	4-14 =	8-17			
9 =	2-12 =	4-15 =	8-18			
10 =	2-13 =	4-16				
11 =	2-14 =	4-17				
12 =	2-15 =	4-18				
13 =	2-16 =	4-19				
14 =	2-17					

TABLE 28

Kilowatts to Amperes Conversion Tables

115 Volts D.C. and A.C. Single Phase—Unity P.F.

K.W.	0	1	2	3	4	5	6	7	8	9
0 ...		8.70	17.39	26.09	34.78	43.48	52.17	60.87	69.56	78.26
0.1 ...	0.87	9.57	18.26	26.95	35.65	44.35	53.04	61.74	70.43	79.13
0.2 ...	1.74	10.43	19.13	27.82	36.52	45.22	53.91	62.61	71.30	80.00
0.3 ...	2.61	11.30	20.00	28.69	37.39	46.09	54.78	63.48	72.17	80.87
0.4 ...	3.48	12.17	20.87	29.56	38.26	46.96	55.65	64.35	73.04	81.74
0.5 ...	4.35	13.04	21.74	30.43	39.13	47.83	56.52	65.22	73.91	82.61
0.6 ...	5.22	13.91	22.61	31.30	40.00	48.75	57.39	66.09	74.78	83.48
0.7 ...	6.09	14.78	23.48	32.17	40.87	49.56	58.26	66.96	75.65	84.35
0.8 ...	6.96	15.65	24.35	33.04	41.74	50.43	59.13	67.83	76.52	85.22
0.9 ...	7.83	16.52	25.22	33.91	42.61	51.30	60.00	68.70	77.39	86.09

Table 28 *(Cont.)*

208 Volts, 3-Phase, Unity P.F.

K.W.	0	1	2	3	4	5	6	7	8	9
0 ...		2.78	5.56	8.34	11.12	13.90	16.67	19.45	22.23	25.01
0.1 ...	.28	3.06	5.84	8.61	11.39	14.17	16.95	19.73	22.51	25.29
0.2 ...	.56	3.33	6.11	8.89	11.67	14.45	17.23	20.01	22.79	25.57
0.3 ...	.83	3.61	6.39	9.17	11.95	14.73	17.51	20.29	23.07	25.84
0.4 ...	1.11	3.89	6.67	9.45	12.23	15.01	17.79	20.56	23.34	26.12
0.5 ...	1.39	4.17	6.95	9.73	12.51	15.28	18.06	20.84	23.62	26.40
0.6 ...	1.67	4.45	7.23	10.00	12.78	15.56	18.34	21.12	23.90	26.68
0.7 ...	1.95	4.72	7.50	10.28	13.06	15.84	18.62	21.40	24.18	26.96
0.8 ...	2.22	5.00	7.78	10.56	13.34	16.12	18.90	21.68	24.46	27.23
0.9 ...	2.50	5.28	8.06	10.84	13.62	16.40	19.18	21.95	24.73	27.51

For power factor different from unity, divide the current given in the Table by the power factor expressed as a decimal.

TABLE 29

Horsepower to Amperes Conversion Tables

220 Volts D-C 90% Efficiency

H.P.	0	10	20	30	40	50	60	70	80	90
0 ...		37.68	75.35	113.0	150.7	188.4	226.1	263.7	301.4	339.1
1 ...	3.77	41.44	79.12	116.8	154.5	191.2	229.8	267.5	305.2	342.9
2 ...	7.54	45.21	82.88	120.6	158.2	195.9	233.6	271.3	309.0	346.3
3 ...	11.30	48.98	86.66	124.3	162.0	199.7	237.4	275.0	312.7	350.4
4 ...	15.07	52.75	90.42	128.1	165.8	203.5	241.1	278.8	316.5	354.2
5 ...	18.84	56.52	94.19	131.9	169.6	207.2	244.9	282.6	320.2	357.9
6 ...	22.61	60.28	97.96	135.6	173.3	212.0	248.7	286.4	324.0	361.7
7 ...	26.37	64.05	101.7	139.4	177.1	214.8	252.4	290.1	327.8	365.5
8 ...	30.14	67.82	105.5	143.2	180.9	218.5	256.2	293.9	331.6	369.2
9 ...	33.91	71.59	109.3	146.9	184.6	222.3	260.0	297.7	335.3	373.0

Table 29 *(Cont.)*

220 Volts, 3-Phase, 90% Eff. 85% P.F.

H.P.	0	10	20	30	40	50	60	70	80	90
0		25.56	51.12	76.68	102.2	127.8	153.4	178.9	204.5	230.3
1	2.56	28.12	53.68	79.24	104.8	130.4	155.9	181.5	207.0	232.6
2	5.11	30.67	56.23	81.79	107.4	132.9	158.5	184.0	209.6	235.2
3	7.67	33.23	58.79	84.34	109.9	135.5	161.0	186.6	212.1	237.7
4	10.22	35.78	60.34	86.91	112.5	138.0	163.6	189.1	214.7	240.3
5	12.78	38.34	64.90	89.46	115.0	140.6	166.1	191.7	217.3	242.8
6	15.34	40.90	66.46	92.02	117.6	143.1	168.7	194.3	219.8	245.4
7	17.89	43.45	69.01	94.57	120.1	145.7	171.3	196.8	222.4	247.9
8	20.45	46.01	71.57	97.13	122.7	148.2	173.8	199.4	224.9	250.5
9	23.00	48.56	74.12	99.68	125.2	150.8	176.4	201.9	227.5	253.0

208 Volts, 3-Phase, 90% Eff. 85% P.F.

H.P.	0	10	20	30	40	50	60	70	80	90
0 ...		26.96	53.92	80.89	107.8	134.8	161.8	188.7	215.7	242.7
1 ...	2.69	29.66	56.62	83.58	110.5	137.5	164.5	191.4	218.4	245.4
2 ...	5.39	32.35	59.32	86.28	113.2	140.2	167.2	194.1	221.1	248.1
3 ...	8.09	35.05	62.01	88.97	115.6	142.9	169.9	196.8	223.8	250.7
4 ...	10.78	37.75	64.71	91.67	118.6	145.6	172.6	199.5	226.5	253.4
5 ...	13.48	40.44	67.41	94.37	121.3	148.3	175.3	202.2	229.2	256.1
6 ...	16.18	43.14	70.10	97.06	124.0	151.0	177.9	204.9	231.9	258.8
7 ...	18.87	45.84	72.80	99.76	126.7	153.7	180.6	207.6	234.6	261.5
8 ...	21.57	48.53	75.49	102.5	129.4	156.4	183.3	210.3	237.3	264.2
9 ...	24.27	51.23	78.19	105.2	132.1	159.1	186.0	213.0	240.0	266.9

TABLE 30

Recommended Bus Spacings for Different Voltages

Operating Voltage	Distance Between Bus Centers		Minimum Distance Between Live Side and Ground		Minimum Distance Between Opposite Potentials	
	A	B	A	B	A	B
250	1½"	to 2½"	¾"	to 1½"	1"	to 2"
600	2"	to 3"	1"	to 2"	1½"	to 2½"
1,100	4"	to 5"	1½"	to 2½"	2½"	to 3½"
2,300	5"	to 6½"	2"	to 2¾"	2¾"	to 4"
4,000	6"	to 7½"	2¼"	to 3"	3"	to 4½"
6,600	7"	to 8"	2½"	to 3"	3½"	to 4½"
7,500	8"	to 9"	3¾"	to 3¼"	4"	to 4½"
9,000	9"	to 10"	3"	to 3½"	4¼"	to 4½"
11,000	9"	to 11"	3¼"	to 3¾"	4½"	to 4¾"
13,200	9"	to 12"	3½"	to 4¼"	4¾"	to 5"
15,000	9"	to 14"	3¾"	to 4½"	5"	to 5½"
16,500	10"	to 14"	4½"	to 5"	5½'	to 6"
18,000	11"	to 14"	5"	to 6"	6"	to 7"
22,000	12"	to 15"	6"	to 7"	7½"	to 9"
26,000	14"	to 16"	8"	to 9"	10'	to 12"
35,000	18"	to 22"	10"	to 12"	12"	to 15"
45,000	22"	to 27"	13½"	to 15"	16"	to 18"
56,000	28"	to 31"	16"	to 17½"	17½"	to 19"
66,000	34"	to 38"	18½"	to 23"	22"	to 24"
75,000	36"	to 42"	25"	to 27½"	26"	to 30"
90,000	46"	to 54"	27"	to 29"	32"	to 35"
104,000	54"	to 60"	28½"	to 32"	34½"	to 39"
110,000	60"	to 72"	33"	to 36"	[illegible]3"	to 4[illegible]"
122,000	66"	to 78"	35½"	to 39"	42"	to 47"
134,000	74"	to 84"	39"	to 41"	48½"	to 56"
148,000	82"	to 96"	45"	to 50"	59"	to 67"
160,000	88"	to 105"	53"	to 63"	70"	to 85"

The distances given in "A" are based on a safety factor of 3½ times between live parts of opposite polarity and safety factor 3 times between live parts and ground. "B" columns show good practice for larger plants.

TABLE 31

Sizes, Weights and Current Carrying Capacity of Drawn Rectangular Copper Bars

Bar Size	Cross Section, Sq. In.	Weight, lbs. per foot	Carrying Capacity at Current Density of 750 amps.	800 amps.	1000 amps.
2 × 1/8	0.250	0.962	188	200	250
2 1/4 × 1/8	0.281	1.080	211	225	281
2 1/2 × 1/8	0.313	1.205	235	251	313
2 3/4 × 1/8	0.344	1.324	258	275	344
3 × 1/8	0.375	1.444	281	300	375
3 1/2 × 1/8	0.437	1.680	328	350	437
4 × 1/8	0.500	1.925	375	400	500
2 × 1/4	0.500	1.925	375	400	500
2 1/4 × 1/4	0.562	2.165	422	450	562

Table 31 *(Cont.)*

Bar Size	Cross Section, Sq. In.	Weight, lbs. per foot	Carrying Capacity at Current Density of		
			750 amps.	800 amps.	1000 amps.
2½ × ¼	0.625	2.41	469	500	625
2¾ × ¼	0.688	2.65	516	550	688
3 × ¼	0.750	2.89	563	600	750
3½ × ¼	0.875	3.37	656	700	875
4 × ¼	1.000	3.85	750	800	1000
2 × ⅜	0.750	2.89	563	600	750
2¼ × ⅜	0.842	3.24	632	673	842
2½ × ⅜	0.938	3.61	703	751	938
2¾ × ⅜	1.030	3.97	772	824	1030
3 × ⅜	1.125	4.33	844	900	1125
3½ × ⅜	1.314	5.06	985	1052	1314
4 × ⅜	1.500	5.77	1125	1200	1498

TABLE 32 Rigid Steel Conduit

The following Tables 32(a), (b), (c), (d), (e) and (f) give useful information on rigid steel conduit. Table 32(f) is taken from NEC Section 346-10.

TABLE 32(a) Dimensions and Weights of Rigid Steel Conduit

1	2	3	4	5	6
Nominal or Trade Size of Conduit (Ins.)	Inside Diam. (Ins.)	Outside Diam. (Ins.)	Wall Thickness (Ins.)	Length Without Coupling (Ft. & Ins.)	Min. Wt. of 10 Unit Lengths with Couplings Attached (Lbs.)
1/4	0.364	0.540	0.088	9–11 1/2	38.5
3/8	0.493	0.675	0.091	9–11 1/2	51.5
1/2	0.622	0.840	0.109	9–11 1/4	79.0
3/4	0.824	1.050	0.113	9–11 1/4	105.0
1	1.049	1.315	0.133	9–11	153.0
1 1/4	1.380	1.660	0.140	9–11	201.0
1 1/2	1.610	1.900	0.145	9–11	249.0
2	2.067	2.375	0.154	9–11	334.0
2 1/2	2.469	2.875	0.203	9–10 1/2	527.0
3	3.068	3.500	0.216	9–10 1/2	690.0
3 1/2	3.548	4.000	0.226	9–10 1/4	831.0
4	4.026	4.500	0.237	9–10 1/4	982.0
5	5.047	5.563	0.258	9–10	1344.0
6	6.065	6.625	0.280	9–10	1770.0

TABLE 32(b) Dimensions of Threads for Rigid Steel Conduit

1	2	3	3a	4	5
Nominal or Trade Size of Conduit (Inches)	Threads per Inch	Pitch Diam. End of Thread E_0 (Inches)		Length of Thread (Inches)	
		Taper Per Foot		Effective	Overall
		3/4 inch	3/8 inch	L_2	L_4
1/4	18	0.4774	0.4865	0.40	0.59
3/8	18	0.6120	0.6213	0.41	0.60
1/2	14	0.7584	0.7706	0.53	0.78
3/4	14	0.9677	0.9803	0.55	0.79
1	11 1/2	1.2136		0.68	0.98
1 1/4	11 1/2	1.5571		0.71	1.01
1 1/2	11 1/2	1.7961		0.72	1.03
2	11 1/2	2.2690		0.76	1.06
2 1/2	8	2.7195		1.14	1.57
3	8	3.3406		1.20	1.63
3 1/2	8	3.8375		1.25	1.68
4	8	4.3344		1.30	1.73
5	8	5.3907		1.41	1.84
6	8	6.4461		1.51	1.95

*Tables 32(a), (b), (c), (d) and (e) courtesy of American Iron and Steel Institute.

TABLE 32(c)

Dimensions of 90° Elbows and Weights of Nipples per Hundred

1	2	3	4	5
	ELBOWS AND BENDS		NIPPLES	
Nominal or Trade Size of Conduit (Inches)	Minimum Radius to Center of Tube (Inches)	Minimum Straight Length Ls at Each End (Inches)	A	B
1/4				
3/8				
1/2	4	1 1/2	0.065	2
3/4	4 1/2	1 1/2	0.090	4
1	5 3/4	1 7/8	0.125	9
1 1/4	7 1/4	2	0.164	10
1 1/2	8 1/4	2	0.202	11
2	9 1/2	2	0.269	14
2 1/2	10 1/2	3	0.430	60
3	13	3 1/8	0.561	70
3 1/2	15	3 1/4	0.663	90
4	16	3 3/8	0.786	115
5	24	3 5/8	1.060	170
6	30	3 3/4	1.410	200

Each lot of 100 nipples shall weigh not less than the no. of lbs. determined by the formula: $W = 100 \times L \times A - B$

Where: W = wt. of 100 nipples in lbs.;
L = length of one nipple in ins.;
A = wt. of nipple per in. in lbs.;
B = wt. in lbs., lost in threading 100 nipples.

TABLE 32(d)

Dimensions and Weights of Couplings for Rigid Steel Conduit

1	2	3	4
Nominal or Trade Size of Conduit (Inches)	Outside Diameter (Inches)	Length (Inches)	Minimum Weight (Pounds)
1/4	0.719	13/16	.055
3/8	0.875	13/16	.075
1/2	1.010	19/16	.115
3/4	1.250	1 5/8	.170
1	1.525	2	.300
1 1/4	1.869	2 1/16	.370
1 1/2	2.155	2 1/16	.515
2	2.730	2 1/8	.875
2 1/2	3.250	3 1/8	1.675
3	4.000	3 1/4	2.825
3 1/2	4.500	3 3/8	3.400
4	5.000	3 1/2	3.800
5	6.296	3 3/4	7.500
6	7.390	4	9.750

TABLE 32(e)

Range of spacings in inches between centers of conduits to provide clearance between locknuts.

Size	1/2	3/4	1	1 1/4	1 1/2	2	2 1/2	3	3 1/2	4	4 1/2	5	6
1/2 . . .	1 3/16 1 3/8												
3/4 . . .	1 5/16 1 1/2	1 7/16 1 5/8											
1 . . .	1 1/2 1 3/4	1 5/8 1 7/8	1 3/4 2										
1 1/4 . . .	1 3/4 2	1 7/8 2 1/8	2 2 1/4	2 1/4 2 1/2									
1 1/2 . . .	1 15/16 2 1/8	2 1/16 2 1/4	2 3/16 2 3/8	2 7/16 2 5/8	2 9/16 2 3/4								
2 . . .	2 3/16 2 3/8	2 5/16 2 1/2	2 1/2 2 3/4	2 3/4 3	2 7/8 3 1/8	3 1/8 3 3/8							

TABLE 32(e) (Cont.)

Size	1/2	3/4	1	1 1/4	1 1/2	2	2 1/2	3	3 1/2	4	4 1/2	5	6
2 1/2 . . .	2 7/16 2 5/8	2 9/16 2 3/4	2 3/4 3	3 3 1/4	3 1/8 3 3/8	3 3/8 3 5/8	3 5/8 4						
3 . . .	2 13/16 3	2 15/16 3 1/8	3 1/16 3 3/8	3 5/16 3 5/8	3 7/16 3 3/4	3 3/4 4	4 4 3/8	4 5/16 4 3/4					
3 1/2 . . .	3 1/8 3 3/8	3 1/4 3 1/2	3 3/8 3 5/8	3 5/8 3 7/8	3 3/4 4	4 1/16 4 3/8	4 5/16 4 5/8	4 5/8 5	4 15/16 5 3/8				
4 . . .	3 7/16 3 3/4	3 9/16 3 7/8	3 11/16 4	3 15/16 4 1/4	4 1/16 4 3/8	4 3/8 4 3/4	4 5/8 5	4 15/16 5 3/8	5 1/4 5 5/8	5 9/16 6			
4 1/2 . . .	3 3/4 4	3 7/8 4 1/8	4 4 1/4	4 1/4 4 1/2	4 3/8 4 3/4	4 5/8 5	4 7/8 5 1/4	5 1/4 5 5/8	5 9/16 6	5 7/8 6 1/4	6 1/8 6 1/2		
5 . . .	4 1/8 4 3/8	4 1/4 4 1/2	4 3/8 4 5/8	4 5/8 4 7/8	4 3/4 5	5 5 3/8	5 1/4 5 5/8	5 9/16 6	5 7/8 6 1/4	6 3/16 6 5/8	6 1/2 7	6 13/16 7 1/4	
6 . . .	4 3/4 5	4 7/8 5 1/8	5 5 1/4	5 1/4 5 1/2	5 3/8 5 5/8	5 5/8 6	5 7/8 6 1/4	6 3/16 6 5/8	6 1/2 7	6 13/16 7 1/4	7 1/8 7 5/8	7 7/16 8	8 1/8 8 5/8

Approximate Diameters

Locknut	1 1/8	1 3/8	1 11/16	2 3/16	2 7/16	3	3 7/16	4 3/16	4 13/16	5 3/8	6	6 11/16	7 15/16
Bushing	1	1 1/4	1 1/2	1 7/8	2 1/8	2 5/8	3 3/16	3 7/8	4 7/16	5	5 9/16	6 1/4	7 3/8
Conduit	7/8	1 1/16	1 3/8	1 11/16	1 15/16	2 3/8	2 7/8	3 1/2	4	4 1/2	5	5 9/16	6 5/8

Number per 100 Lbs.

Locknuts	5882	3845	2500	1250	833	625	500	333	278	208	133	111	53
Bushings	4166	2858	1110	833	667	500	333	250	179	125	83	67	43.5

TABLE 32(f)
Radius of Conduit Bends

Section 3470 of the NEC prescribes that the radius of the curve of the inner edge of any field bend shall not be less than shown in the following table:

Size of Conduit	Conductors Without Lead Sheath	Conductors With Lead Sheath
½ in.	4 in.	6 in.
¾ in.	5 in.	8 in.
1 in.	6 in.	11 in.
1¼ in.	8 in.	14 in.
1½ in.	10 in.	16 in.
2 in.	12 in.	21 in.
2½ in.	15 in.	25 in.
3 in.	18 in.	31 in.
3½ in.	21 in.	36 in.
4 in.	24 in.	40 in.
5 in.	30 in.	50 in.
6 in.	36 in.	61 in.

TABLE 33
Weights of Cabinets and Boxes

Size (inches)	Gauge Steel	Approx. Wt. (lbs.)
12 × 12 × 4	14	12
12 × 18 × 4	14	17
12 × 18 × 8	14	22
12 × 24 × 4	14	21
12 × 24 × 8	14	28
18 × 24 × 4	14	29
18 × 24 × 8	14	37
18 × 30 × 4	14	35
18 × 30 × 8	14	44
18 × 30 × 12	14	53
24 × 24 × 4	14	37
24 × 24 × 8	14	46
24 × 24 × 12	14	55
24 × 30 × 4	14	45
24 × 30 × 8	14	54
24 × 30 × 12	14	64
24 × 36 × 8	12	88
24 × 36 × 12	12	104
24 × 42 × 8	12	100
24 × 42 × 12	12	117
30 × 30 × 8	12	90
30 × 30 × 12	12	106
30 × 36 × 8	12	105
30 × 36 × 12	12	122
36 × 36 × 8	12	122
36 × 36 × 12	12	140
36 × 36 × 16	12	158
36 × 42 × 8	10	178
36 × 42 × 12	10	204
36 × 42 × 16	10	230
36 × 48 × 8	10	200
36 × 48 × 12	10	230
36 × 48 × 16	10	260
42 × 42 × 8	10	194
42 × 42 × 12	10	221
42 × 42 × 16	10	248
42 × 48 × 8	10	227
42 × 48 × 12	10	256
42 × 48 × 16	10	286

TABLE 33 (Cont.)

Size (inches)	Gauge Steel	Approx. Wt. (lbs.)
42 × 60 × 12	10	310
42 × 60 × 16	10	348
42 × 60 × 20	10	382
48 × 48 × 12	10	285
48 × 48 × 16	10	317
48 × 48 × 20	10	348
48 × 60 × 16	10	379
48 × 60 × 20	10	415
48 × 72 × 16	10	440
48 × 72 × 20	10	480
60 × 60 × 20	10	475
60 × 60 × 24	10	515
60 × 72 × 20	8	675
60 × 72 × 24	8	715

Weights per square foot of steel:
No. 14—3.2 lbs., No. 12—4.5 lbs., No. 10—5.75 lbs.

TABLE 34—Weights of A-C Motors (Motor only—pounds)

208, 220, 440 or 550 volts, 60 cycle, 3 phase

H.P.	3500 rpm (2-pole)	1750 rpm (4-pole)	1150 rpm (6-pole)	875 rpm (8-pole)
1		70	80	115
5	115	155	180	240
10	180	240	265	450
25	450	450	620	650
50	620	650	970	1100
100	970	1100	1700	2500
150	1300	1700	2500	2500
200	1700	2500	2500	3100
2300 volts, 60 cycle, 3 phase				
50		970	1100	1300
100	1100	1300	2500	2500
150	1700	2800	2800	2800
200	2800	2800	2800	3800

TABLE 35—Weights of D-C Motors (Motor only—pounds)

115, 230 volts, constant speed

H.P.	3500 rpm	1750 rpm	1150 rpm	850 rpm	690 rpm	575 rpm
1		85	85	155	165	195
5	165	195	255	475	500	500
10	255	475	545	785	785	880
25	585	785	980	1020	1320	1555
50	850	1020	1585	1945	2280	2675
75		1550	2280	2675	2675	2925
100		1930	2675	2925	2925	3725
150		2675	3675	3840	3840	4860
200		3150	3870	4985	5950	6800

TABLE 36—Transformer Weights—Single Phase, 60 Cycle (pounds)

440, 550 or 600 volts to 115/230 volts

Kva	Air-Cooled*	Oil-Cooled
3	85	155
5	120	180
7.5	155	
10	195	275
15	225	330
25	395	460
37.5	515	670
50	665	820
75	940	1115
100	1400	1250

*3 to 50 kva, Class H Insulation, 75 and 100 kva, Class B Insulation.

TABLE 37
Typical Data on Light Sources

INCANDESCENT LAMPS

Lamp Designation	Rated Life (Hours)	Initial Output (Lumens)	Mean Output (Lumens)	Burnouts Per BLH●
100-watt A21	750	1,630	–	–
150-watt A23	750	2,700	–	–
200-watt A25	750	3,700	3,500	381
300-watt PS35	1,000	5,650	5,100	196
500-watt PS40*	1,000	11,200	10,200	98
750-watt PS52*	1,000	17,100	15,800	63
1000-watt 5S52*	1,000	23,400	21,300	47
500-watt R52	2,000	7,500	6,900	73
750-watt R52	2,000	12,500	10,000	50

*Coiled-coil filament axially mounted.

FLUORESCENT LAMPS

Lamp Designation	Rated Life (Hours)*	Initial Output (Lumens)	Mean** Output (Lumens)	Burnouts Per BLH●
F40T12/CW	7,500	2,500	2,150	62
F40T12/D	7,500	2,300	2,000	67
F90T17/CW	7,500	5,150	4,250	31
F90T17/D	7,500	4,800	4,000	33
F48T12/CW/RS (800 ma)	7,500	3,100	–	–
F72T12/CW/RS (800 ma)	7,500	4,800	–	–
F96T12/CW/RS (800 ma)	7,500	7,250	6,150	22
F48T12/CW Slimline	7,500	2,300	2,000	67
F96T12/CW Slimline	7,500	5,050	4,500	30

*Based on 3 burning hours per start.
**Approximate mean output at 40% life.

MERCURY VAPOR LAMPS

Lamp Designation	Rated Life* (Hours)	Initial Output (Lumens)	Mean Output (Lumens)	Burnouts Per BLH●
H250A5	6,000	11,000	8,900	18.7
H400E1	6,000**	21,000	16,400	10.2
H400J1	6,000**	20,000	15,600	11.7
H400R1	6,000**	18,000	14,900	11.2
H400RC1	6,000**	20,500	17,000	9.8
H1000A15	6,000	54,000	35,100	4.8
H3000A9	6,000	132,000	103,000	1.6

*Rated life data shown are averages from tests of lamps burning five hours per start in appropriate industrial type reflectors.
**Designates economic life of the lamp under typical average operating conditions.
●Average lamp burnouts per billion lumen hours (calculated).

TABLE 37 (Cont.)

Number and Type of Lamps	Weight (lbs)	Line Current (amps)	Watts Consumed: Ballast	Watts Consumed: Per Lamp*	Lumens per Lamp	Lumens per Watt**
FOR 60 CYCLE, 110-125-VOLT CIRCUITS						
2–20	3.4	.42	9	24.5	1000	40.8
2–30	6.2	.70	18	39.0	1890	48.5
2–40	5.5	.85	16	48.0	2500	52.1
2–40T17IS	9.5	.95	30	55.0	2500	45.4
2–48T12	9.5	.95	30	53.0	2300	43.4
2–72T12	13.8	1.60	34	72.0	3600	50.0
2–96T12	13.8	1.60	34	91.0	5050	55.4
2–48T12RS	4.0	.85	12	46.0	2500	54.4
2–48T12RS/800	10.0	1.30	30	75.0	3100	41.4
2–72T12RS/800	14.0	1.70	30	100.0	4800	48.0
2–96T12RS/800	14.0	2.00	20	115.0	7250	63.0
2–90T17	10.0	1.00	21	100.5	5150	51.2
FOR 60 CYCLE, 220-250-VOLT CIRCUITS						
2–40	5.7	.45	18	49.0	2500	51.0
2–90T17	14.3	.93	31	105.5	5150	48.8
2–48T12RS	5.5	.43	13	46.5	2500	53.8
FOR 60 CYCLE, 240-280-VOLT CIRCUITS						
2–40	5.3	.37	18	49.0	2500	51.0
2–48T12	10.0	.43	28	52.0	2300	49.2
2–72T12	9.5	.74	30	70.0	3600	51.4
2–96T12	9.5	.74	30	89.0	5050	56.8
2–48T12RS	4.0	.39	12	46.0	2500	54.3
3–48T12RS	11.0	.55	18	46.0	2500	54.3
2–96T12RS/800	14.0	.88	20	115.0	7250	63.0
2–90T17	14.3	.83	31	105.5	5150	48.7
4–90T17	16.5	1.65	36	99.0	5150	52.0

*Includes lamp wattage plus watts loss in ballast.
**Overall efficiency, including ballast watts loss.

TABLE 38
Typical Wattage Loads of Domestic Appliances

Appliance	Typical Load in Watts
Air Conditioner–3/4 hp	1200
" " –1 1/2 hp	2400
Dishwasher	1200
Broiler	1500
Coffee Maker	1000
Freezer	350
Fryer	1300
Ironer	1650
Washing Machine	1200
Dryer	5000
Range	12000
Built In Oven	4500
Range Top	3000
Refrigerator	300
Waste Disposer	300
Water Heater	3000
Hand Iron	1000
Portable Heater	1300
Television	300

TABLE 39 (220-2a)
General Lighting Loads by Occupancies

Type of Occupancy	Unit Load per Sq. Ft. (Watts)
Armories and Auditoriums	1
Banks	5
Barber Shops and Beauty Parlors	3
Churches	1
Clubs	2
Court Rooms	2
*Dwellings (Other Than Hotels)	3
Garages—Commercial (storage)	½
Hospitals	2
*Hotels and Motels, including apartment houses without provisions for cooking by tenants	2
Industrial Commercial (Loft) Buildings	2
Lodge Rooms	1½
Office Buildings	5
Restaurants	2
Schools	3
Stores	3
Warehouses Storage	¼
In any of the above occupancies except single-family dwellings and individual apartments of multi-family dwellings:	
Assembly Halls and Auditoriums	1
Halls, Corridors, Closets	½
Storage Spaces	¼

*All receptacle outlets of 15-ampere or less rating in single-family and multi-family dwellings and in guest rooms of hotels and motels [except those connected to the receptacle circuits specified in Section 220-3(b)] may be considered as outlets for general illumination, and no additional load need be included for such outlets.

TABLE 40a (250-94a)—Grounding Electrode Conductor for Grounded Systems

Size of Largest Service-Entrance Conductor or Equivalent for Parallel Conductors		Size of Grounding Electrode Conductor	
Copper	Aluminum or Copper-Clad Aluminum	Copper	*Aluminum or Copper-Clad Aluminum
2 or smaller	0 or smaller	8	6
1 or 0	2/0 or 3/0	6	4
2/0 or 3/0	4/0 or 250 MCM	4	2
Over 3/0 thru 350 MCM	Over 250 MCM thru 500 MCM	2	0
Over 350 MCM	Over 500 MCM	0	3/0

thru 600 MCM	thru 900 MCM		
Over 600 MCM thru 1100 MCM	Over 900 MCM thru 1750 MCM	2/0	4/0
Over 1100 MCM	Over 1750 MCM	3/0	250 MCM

Where there are no service-entrance conductors, the grounding electrode conductor size shall be determined by the equivalent size of the largest service-entrance conductor required for the load to be served.

*See installation restrictions in Section 250-92(a) of 1971 NEC.

See Section 250-23(b) of 1971 NEC.

TABLE 40b (250-94b)—Grounding Electrode Conductor of Ungrounded Systems

Size of Largest Service-Entrance Conductor or Equivalent for Parallel Conductors		Size of Grounding Electrode Conductor			
Copper	Aluminum or Copper-Clad Aluminum	Copper	*Aluminum or Copper-Clad Aluminum	Conduit or Pipe	Electrical Metallic Tubing
2 or smaller	0 or smaller	8	6	½	½
1 or 0	2/0 or 3/0	6	4	½	1

2/0 or 3/0	4/0 or 250 MCM	4	2	¾	1¼
Over 3/0 thru 350 MCM	Over 250 MCM thru 500 MCM	2	0	¾	1¼
Over 350 MCM thru 600 MCM	Over 500 MCM thru 900 MCM	0	3/0	1	2
Over 600 MCM thru 1100 MCM	Over 900 MCM thru 1750 MCM	2/0	4/0	1	2
Over 1100 MCM	Over 1750 MCM	3/0	250 MCM	1	2

Where there are no service-entrance conductors the grounding electrode conductor size shall be determined by the equivalent size of the largest service-entrance conductor required for the load to be served.

*See installation restrictions in Section 250-92(a) of 1971 NEC.

TABLE 40c (250-95)
Size of Equipment Grounding Conductors for Grounding Interior Raceway and Equipment

Rating or Setting of Automatic Overcurrent Device in Circuit Ahead of Equipment, Conduit, etc., Not Exceeding (Amperes)	Size	
	Copper Wire No.	Aluminum or Copper-Clad Aluminum Wire No.*
15	14	12
20	12	10
30	10	8
40	10	8
60	10	8
100	8	6
200	6	4
400	3	1
600	1	2/0
800	0	3/0
1000	2/0	4/0
1200	3/0	250 MCM
1600	4/0	350 "
2000	250 MCM	400 "
2500	350 "	500 "
3000	400 "	600 "
4000	500 "	800 "
5000	700 "	1000 "
6000	800 "	1200 "

*See installation restrictions in Section 250-92(a) of 1971 NEC.

TABLE 41
Drill Sizes For Machine Screws

Fractional Size Drills Inches	Wire Gauge and Letter Size Drills	Decimal Equivalent Inches	Machine Screw No.	Machine Screw O. D.	Machine Screw Threads Per Inch	Machine Screw Tap Drill	Machine Screw Clear. Drill
	80	.0135					
	79	.0145					
1/64		.0156					
	78	.0160					
	77	.0180					
	76	.0200					
	75	.0210					
	74	.0225					
	73	.0240					
	72	.0250					
	71	.0260					
	70	.0280					
	69	.0292					
	68	.0310					
1/32		.0312					
	67	.0320					
	66	.0330					
	65	.0350					
	64	.0360					
	63	.0370					
	62	.0380					
	61	.0390					
	60	.0400					
	59	.0410					
	58	.0420					
	57	.0430					
	56	.0465					
3/64		.0469					
	55	.0520					
	54	.0550					
	53	.0595					
1/16		.0625	0	.060	80	3/64	1/16
	52	.0635					
	51	.0670					
	50	.0700					
	49	.0730					
	48	.0760					
5/64		.0781			72	53	47
	47	.0785	1	.078	64	53	47
	46	.0810					
	45	.0820					
	44	.0860					
	43	.0890			64	50	42
	42	.0935	2	.086	56	50	42
3/32		.0937					
	41	.0960					
	40	.0980					
	39	.0995					
	38	.1015			48	47	37
	37	.1040	3	.099	56	45	37
	36	.1065					

TABLE 41 (Cont.)

Fractional Size Drills Inches	Wire Gauge and Letter Size Drills	Decimal Equivalent Inches	Machine Screw No.	O. D.	Threads Per Inch	Tap Drill	Clear. Drill
7/64		.1094					
	35	.1100					
	34	.1110					
	33	.1130					
	32	.1160			36	44	31
	31	.1200	4	.112	40	43	31
1/8		.1250			48	42	31
	30	.1285					
	29	.1360	5	.125	40	38	29
	28	.1405			44	37	29
9/64		.1406			32	36	27
	27	.1440	6	.138	32	36	27
	26	.1470			40	33	27
	25	.1495					
	24	.1520					
	23	.1540					
5/32		.1562					
	22	.1570					
	21	.1590					
	20	.1610					
	19	.1660					
	18	.1695	8	.164	32	29	18
11/64		.1719			36	29	18
	17	.1730					
	16	.1770					
	15	.1800					
	14	.1820					
	13	.1850					
3/16		.1875					
	12	.1890					
	11	.1910					
	10	.1935					
	9	.1960	10	.190	24	25	9
	8	.1990			32	21	9
	7	.2010					
13/64		.2031					
	6	.2040					
	5	.2055					
	4	.2090					
	3	.2130					
7/32		.2187					
	2	.2210					
	1	.2280	12	.216	24	16	1
	A	.2340			28	14	1
15/64		.2344					
	B	.2380					
	C	.2420					
	D	.2460					
1/4	E	.2500					
	F	.2570					
	G	.2610					

TABLE 41 (Cont.)

Fractional Size Drills Inches	Wire Gauge and Letter Size Drills	Decimal Equivalent Inches	Machine Screw				
			No.	O. D.	Threads Per Inch	Tap Drill	Clear. Drill
17/64		.2656	1/4	.250	20	7	17/64
	H	.2660			28	3	17/64
	I	.2720					
	J	.2770					
	K	.2810					
9/32		.2812					
	L	.2900					
	M	.2950					
19/64		.2969					
	N	.3020					
5/16		.3125					
	O	.3160					
	P	.3230					
21/64		.3281	5/16	.3125	18	F	21/64
	Q	.3320			24	I	21/64
	R	.3390					
11/32		.3437					
	S	.3480					
	T	.3580					
23/64		.3594					
	U	.3680					
3/8		.3750					
	V	.3770					
	W	.3860					
25/64		.3906	3/8	.375	16	5/16	25/64
	X	.3970			24	Q	25/64
	Y	.4040					
13/32		.4062					
	Z	.4130					
27/64		.4219					
7/16		.4375					
29/64		.4531	7/16	.4515	14	U	29/64
15/32		.4687			20	25/64	29/64
31/64		.4844					
1/2		.5000					
33/64		.5156	1/2	.500	13	27/64	33/64
					20	29/64	33/64

TABLE 42—Gauges for Wire, Sheet, and Twist Drills

No. of Gauge	Washburn & Moen	American or Brown & Sharpe	Birmingham or Stubbs	U. S. standard for Plate (Iron & Steel)	Stubbs Steel Wire	Imperial Wire Gauge	Morse Twist Drill and Steel Wire	Wood and Machine Screws	American S. & W. Piano & Music Wire
7–0	.....	.....	....	.500	....	.....	.....	....	....
6–0	.....	.....	....	.469	....	.464	.....	....	....
5–0	.....	.....	....	.438	....	.432	.....	....	.005
4–0	.394	.460	.454	.406	....	.400	.....	....	.006
3–0	.362	.410	.425	.375	....	.372	.....	.032	.007
2–0	.331	.365	.380	.344	....	.348	.....	.045	.008
0	.307	.325	.340	.313	....	.324	.....	.058	.009
1	.283	.289	.300	.281	.227	.300	.228	.071	.010
2	.263	.258	.284	.266	.219	.276	.221	.084	.011
3	.244	.229	.259	.250	.212	.252	.213	.097	.012
4	.225	.204	.238	.234	.207	.232	.209	.110	.013
5	.207	.182	.220	.219	.204	.212	.206	.124	.014
6	.192	.162	.203	.203	.201	.192	.204	.137	.016
7	.177	.144	.180	.188	.199	.176	.201	.150	.018
8	.162	.128	.165	.172	.197	.160	.199	.163	.020
9	.148	.114	.148	.156	.194	.144	.196	.176	.022

10	.135	.102	.134	.141	.191	.128	.194	.189	.024
11	.120	.091	.120	.125	.188	.116	.191	.203	.026
12	.105	.081	.109	.109	.185	.104	.189	.216	.029
13	.092	.072	.095	.094	.182	.092	.185	.229	.031
14	.080	.064	.083	.078	.180	.080	.182	.242	.033
15	.072	.057	.072	.070	.178	.072	.180	.255	.035
16	.063	.051	.065	.063	.175	.064	.177	.268	.037
17	.054	.045	.058	.056	.172	.056	.173	.282	.039
18	.047	.040	.049	.050	.168	.048	.170	.295	.041
19	.041	.036	.042	.044	.164	.040	.166	.308	.043
20	.035	.032	.035	.038	.161	.036	.161	.321	.045
21	.032	.028	.032	.034	.157	.032	.159	.334	.047
22	.028	.025	.028	.031	.155	.028	.157	.347	.049
23	.025	.023	.025	.028	.153	.024	.154	.360	.051
24	.023	.020	.022	.025	.151	.022	.152	.374	.055
25	.020	.018	.020	.022	.148	.020	.150	.387	.059
26	.018	.016	.018	.019	.146	.018	.147	.400	.063
27	.0173	.0141	.016	.0171	.143	.0164	.144	.413	.067
28	.0162	.0126	.014	.0156	.139	.0149	.141	.426	.071
29	.015	.0112	.013	.014	.134	.0136	.136	.439	.075
30	.014	.010	.012	.0125	.127	.0124	.129	.453	.080
31	.0132	.0089	.010	.0109	.120	.0116	.120	.466	.085
32	.0128	.0079	.009	.0101	.115	.0108	.116	.479	.090
33	.0118	.007	.008	.0093	.112	.010	.113	.492	.095
34	.0104	.0063	.007	.0085	.110	.0092	.111	.505	.100
35	.0095	.0056	.005	.0078	.108	.0084	.110	.518	.106

TABLE 42 *(Cont.)*

No. of Gauge	Washburn & Moen	American or Brown & Sharpe	Birmingham or Stubbs	U. S. standard for Plate (Iron & Steel)	Stubbs Steel Wire	Imperial Wire Gauge	Morse Twist Drill and Steel Wire	Wood and Machine Screws	American S. & W. Piano & Music Wire
36	.009	.005	.004	.007	.106	.0076	.1065	.532	.112
37	.....	.0044	....	.0066	.103	.0068	.104	.545	.118
38	.....	.0039	....	.0062	.101	.006	.1015	.558	.124
39	.....	.0035	....	.....	.099	.0052	.0995	.571	.130
40	.....	.0031	....	.....	.097	.0048	.098	.584	.138
41	.....	.....	....	.....	.095	.....	.096	.597	.146
42	.....	.....	....	.....	.092	.....	.094	.611	.154
43	.....	.....	....	.....	.088	.....	.089	.624	.162
44	.....	.....	....	.....	.085	.....	.086	.637	.170
45	.....	.....	....	.....	.081	.....	.082	.650	.180
46	.....	.....	....	.....	.079	.....	.081	.663	....
47	.....	.....	....	.....	.077	.....	.079	.676	....
48	.....	.....	....	.....	.075	.....	.076	.690	....
49	.....	.....	....	.....	.072	.....	.073	.703	....
50	.....	.....	....	.....	.069	.....	.070	.716	....

TABLE 43
Fraction and Decimal Equivalents

Fraction	Decimal	Fraction	Decimal
1/64	.015625	33/64	.515625
1/32	.03125	17/32	.53125
3/64	.046875	35/64	.546875
1/16	.0625	9/16	.5625
5/64	.078125	37/64	.578125
3/32	.09375	19/32	.59375
7/64	.109375	39/64	.609375
1-8	.125	5-8	.625
9/64	.140625	41/64	.640625
5/32	.15625	21/32	.65625
11/64	.171875	43/64	.671875
3/16	.1875	11/16	.6875
13/64	.203125	45/64	.703125
7/32	.21875	23/32	.71875
15/64	.234375	47/64	.734375
1-4	.25	3-4	.75
17/64	.265625	49/64	.765625
9/32	.28125	25/32	.78125
19/64	.296875	51/64	.796875
5/16	.3125	13/16	.8125
21/64	.328125	53/64	.828125

TABLE 43 (Cont.)

11/32	.34375	27/32	.84375
23/64	.359375	55/64	.859375
3-8	.375	7-8	.875
25/64	.390625	57/64	.890625
13/32	.40625	29/32	.90625
27/64	.421875	59/64	.921875
7/16	.4375	15/16	.9375
29/64	.453125	61/64	.953125
15/32	.46875	31/32	.96875
31/64	.484375	63/64	.984375
1-2	.5	1	1.0

TABLE 44
Thermometer Scale Conversion

Centigrade—Fahrenheit

Degrees C = 5/9 (F − 32) Degrees F = 9/5 C + 32

°C.	°F.	°C.	°F.	°C.	°F.	°C.	°F.	°C.	°F.
−80	−112.	1	33.8	31	87.8	61	141.8	91	195.8
−70	− 94.	2	35.6	32	89.6	62	143.6	92	197.6
−60	− 65.	3	37.4	33	91.4	63	145.4	93	199.4
−50	− 58.0	4	39.2	34	93.2	64	147.2	94	201.2
−45	− 49.1	5	41.0	35	95.0	65	149.0	95	203.0
−40	− 40.0	6	42.8	36	96.8	66	150.8	96	204.8
−35	− 31.0	7	44.6	37	98.6	67	152.6	97	206.6
−30	− 22.0	8	46.4	38	100.4	68	154.4	98	208.4
−25	− 13.0	9	48.2	39	102.2	69	156.2	99	210.2
−20	− 4.0	10	50.0	40	104.0	70	158.0	100	212.0
−19	− 2.2	11	51.8	41	105.8	71	159.8	105	221.
−18	− .4	12	53.6	42	107.6	72	161.6	110	230.
−17	− 1.4	13	55.4	43	109.4	73	163.4	115	239.
−16	3.2	14	57.2	44	111.2	74	165.2	120	248.
−15	5.0	15	59.0	45	113.0	75	167.0	130	266.
−14	6.8	16	60.8	46	114.8	76	168.8	140	284.
−13	8.6	17	62.6	47	116.6	77	170.6	150	302.
−12	10.4	18	64.4	48	118.4	78	172.4	160	320.
−11	12.2	19	66.2	49	120.2	79	174.2	170	338
−10	14.0	20	68.0	50	122.0	80	176.0	180	356.
− 9	15.8	21	69.8	51	123.8	81	177.8	190	374
− 8	17.6	22	71.6	52	125.6	82	179.6	200	392.
− 7	19.4	23	73.4	53	127.4	83	181.4	250	482.
− 6	21.2	24	75.2	54	129.2	84	183.2	300	572.
− 5	23.0	25	77.0	55	131.0	85	185.0	350	662.
− 4	24.8	26	78.8	56	132.8	86	186.8	400	752.
− 3	26.6	27	80.6	57	134.6	87	188.6	500	932.
− 2	28.4	28	82.4	58	136.4	88	190.4	600	1112.
− 1	30.2	29	84.2	59	138.2	89	192.2	700	1292.
0	32.0	30	86.0	60	140.0	90	194.0	800	1472.
								900	1652.
								1000	1832.

TABLE 45
Conversion Equivalents

Distance

1 inch = 2.54 centimeters
1 kilometer = 0.6214 miles
Circumference of circle = πD (π – 3.1416, D = diameter)
Circumference = 2π radians
1 radian = the arc whose length is equal to the radius = 57.3° of angle

Area

1 sq inch = 6.452 sq centimeters
Area of sphere = πD^2
Area of triangle = 1/2 altitude × base
Area of circle = $\frac{\pi D^2}{4} = 0.7854\ D^2$
Area of trapezoid = 1/2 (sum of parallel sides) × altitude

Volume

1 cubic inch = 16.39 cubic cm = 0.01639 liters
1 cubic yard = 0.7645 cubic meters
1 cubic meter = 35.32 cubic ft
Volume of sphere = $\frac{\pi D^3}{6} = 0.5236\ D^3$
Volume of cone = area of base × 1/3 altitude = $1.0472\ r^2h$

Energy—Power

1 watt = 0.00134 horsepower
1 kilowatthour = 3414 Btu
1 joule = 1 watt-second = 0.738 ft-lb
1 horsepower-hour = 2544 Btu
1 ft-lb per second = 0.001818 horsepower
1 horsepower = 746 watts = 33,000 ft-lb per min = 550 ft-lb per second

Heat—Temperature

1 Btu = 778 ft-lb = 0.00039 hp-hr
1 Btu per min = 17.6 watts
1 Btu = 252 calories (gram)
1 calorie = 0.003968 Btu
1 Btu = heat required to raise 1 pound water 1 degree F
1 calorie = heat required to raise 1 gram water 1 degree C
Degrees Centigrade = 5/9 (°F – 32)
Degrees Fahrenheit = 9/5 (°C) + 32
Degrees Kelvin = °C + 273

Light

1 lux = 1 lumen per square meter = 0.0929 footcandles
1 candle per sq in. = 452 footlamberts
1 lambert = 929 footlamberts = 2.054 candles per sq in.
1 stilb = 1 candle per sq cm = 6.452 candles per sq in.
Mean spherical candlepower = $\frac{\text{lumens}}{12.57}$

Miscellaneous

1 in. mercury = 0.491 lb per sq in.
1 atmosphere = 14.69 lb per sq in. = 29.92 in. Hg = 33.9 ft water
1 year = 8760 hours (365 days)

TABLE 46
Circumference of Circles

Diam.	Circum-ference	Diam.	Circum-ference	Diam.	Circum-ference	Diam.	Circum-ference
1/8	.3927	10	31.41	30	94.24	65	204.2
1/4	.7854	1/2	32.98	31	97.38	66	207.3
3/8	1.178	11	34.55	32	100.5	67	210.4
1/2	1.570	1/2	36.12	33	103.6	68	213.6
5/8	1.963	12	37.69	34	106.8	69	216.7
3/4	2.356	1/2	39.27	35	109.9	70	219.9
7/8	2.748	13	40.84	36	113.0	71	223.0
1	3.141	1/2	42.41	37	116.2	72	226.1
1/8	3.534	14	43.98	38	119.3	73	229.3
1/4	3.927	1/2	45.55	39	122.5	74	232.4
3/8	4.319	15	47.12	40	125.6	75	235.6
1/2	4.712	1/2	48.69	41	128.8	76	238.7
5/8	5.105	16	50.26	42	131.9	77	241.9
3/4	5.497	1/2	51.83	43	135.0	78	245.0
7/8	5.890	17	53.40	44	138.2	79	248.1
2	6.283	1/2	54.97	45	141.3	80	251.3
1/4	7.068	18	56.54	46	144.5	81	254.4
1/2	7.854	1/2	58.11	47	147.6	82	257.6
3/4	8.639	19	59.69	48	150.7	83	260.7
3	9.424	1/2	61.26	49	153.9	84	263.8
1/4	10.21	20	62.83	50	157.0	85	267.0
1/2	10.99	1/2	64.40	51	160.2	86	270.1
3/4	11.78	21	65.97	52	163.3	87	273.3
4	12.56	1/2	67.54	53	166.5	88	276.4
1/2	14.13	22	69.11	54	169.6	89	279.6
5	15.70	1/2	70.68	55	172.7	90	282.7
1/2	17.27	23	72.25	56	175.9	91	285.8
6	18.84	1/2	73.82	57	179.0	92	289.0
1/2	20.42	24	75.39	58	182.2	93	292.1
7	21.99	1/2	76.96	59	185.3	94	295.3
1/2	23.56	25	78.54	60	188.4	95	298.4
8	25.13	26	81.68	61	191.6	96	301.5
1/2	26.70	27	84.82	62	194.7	97	304.7
9	28.27	28	87.96	63	197.9	98	307.8
1/2	29.84	29	91.10	64	201.0	99	311.0

TABLE 47
Area of Circles

Diam.	Area	Diam.	Area	Diam.	Area	Diam.	Area
1/8	0.0123	10	78.54	30	706.86	65	3318.3
1/4	0.0491	1/2	86.59	31	754.76	66	3421.2
3/8	0.1104	11	95.03	32	804.24	67	3525.6
1/2	0.1963	1/2	103.86	33	855.30	68	3631.6
5/8	0.3068	12	113.09	34	907.92	69	3739.2
3/4	0.4418	1/2	122.71	35	962.11	70	3848.4
7/8	0.6013	13	132.73	36	1017.8	71	3959.2
1	0.7854	1/2	143.13	37	1075.2	72	4071.5
1/8	0.9940	14	153.93	38	1134.1	73	4185.4
1/4	1.227	1/2	165.13	39	1194.5	74	4300.8
3/8	1.484	15	176.71	40	1256.6	75	4417.8
1/2	1.767	1/2	188.69	41	1320.2	76	4536.4
5/8	2.073	16	201.06	42	1385.4	77	4656.6
3/4	2.405	1/2	213.82	43	1452.2	78	4778.3
7/8	2.761	17	226.98	44	1520.5	79	4901.6
2	3.141	1/2	240.52	45	1590.4	80	5026.5
1/4	3.976	18	254.45	46	1661.9	81	5153.0
1/2	4.908	1/2	268.80	47	1734.9	82	5281.0
3/4	5.939	19	283.52	48	1809.5	83	5410.6
3	7.068	1/2	298.64	49	1885.7	84	5541.7
1/4	8.295	20	314.16	50	1963.5	85	5674.5
1/2	9.621	1/2	330.06	51	2042.8	86	5808.8
3/4	11.044	21	346.36	52	2123.7	87	5944.6
4	12.566	1/2	363.05	53	2206.1	88	6082.1
1/2	15.904	22	380.13	54	2290.2	89	6221.1
5	19.635	1/2	397.60	55	2375.8	90	6361.7
1/2	23.758	23	415.47	56	2463.0	91	6503.9
6	28.274	1/2	433.73	57	2551.7	92	6647.6
1/2	33.183	24	452.39	58	2642.0	93	6792.9
7	38.484	1/2	471.43	59	2733.9	94	6939.8
1/2	44.178	25	490.87	60	2827.4	95	7088.2
8	50.265	26	530.93	61	2922.4	96	7238.2
1/2	56.745	27	572.55	62	3019.0	97	7389.8
9	63.617	28	615.75	63	3117.2	98	7542.9
1/2	70.882	29	660.52	64	3216.9	99	7697.7

BASIC RELATIONS FOR ELECTRIC CIRCUITS

Direct Current

Ohm's Law:

$$V = IR; \quad I = \frac{V}{R}; \quad R = \frac{V}{I};$$

In direct-current circuits, the electrical power is equal to the product of the voltage and the current:

$$P = VI; \quad \text{or } P = \frac{V^2}{R}; \quad \text{or } P = I^2R$$

where the symbols represent:

P = power in watts
V = impressed voltage in volts
I = current in amperes
R = resistance in ohms.

When several resistances are connected in series, the total resistance is the sum of the separate resistances:

$$R = R_1 + R_2 + R_3 + \ldots.$$

When two resistances are connected in parallel, the total or equivalent resistance is equal to the product of the two resistances divided by their sum:

$$R = \frac{R_1 R_2}{R_1 + R_2}$$

This relation is a special case of the more general one when several resistances are connected in parallel, for which the total or equivalent resistance is found from:

$$\frac{1}{R} = \frac{1}{R_1} + \frac{1}{R_2} + \frac{1}{R_3} + \ldots .$$

Alternating Current

In alternating-current circuits, the opposition to current flow is known as the impedance, which depends upon not only the resistance, but also the inductance and the capacitance which may be present in the circuit. Impedance is related to these three quantities by:

$$X_L = 2\pi f L; \quad X_C = \frac{1}{2\pi f C};$$

$$Z = \sqrt{R^2 + (X_L - X_C)^2}$$

where the symbols represent:

X_L = inductive reactance in ohms
X_C = capacitive reactance in ohms
Z = impedance in ohms
L = inductance in henrys
C = capacitance in farads
f = frequency in cycles per second
$\pi = 3.1416$

Ohm's Law for A-C Circuits:

$$V = IZ; \quad I = \frac{V}{Z}; \quad Z = \frac{V}{I}$$

The impedance for several of the more common circuit arrangements of resistance, inductance, and capacitance are tabulated in the figures shown on next page.

$$Z = R$$

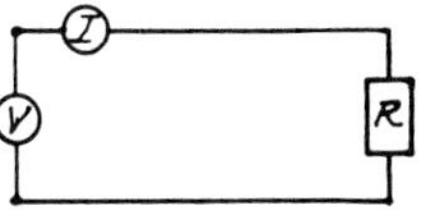

$$Z = \sqrt{R^2 + X_L^2}$$

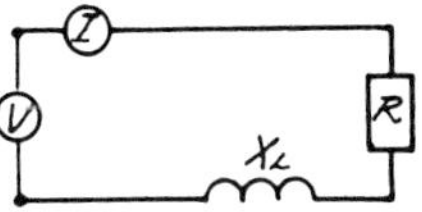

$$Z = \sqrt{R^2 + X_C^2}$$

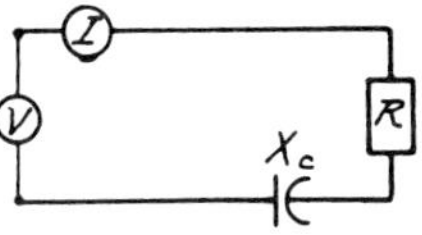

$$Z = \sqrt{R^2 + (X_L - X_C)^2}$$

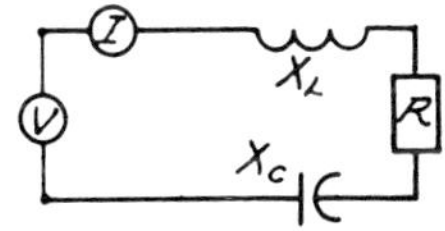

$$Z = \frac{R X_L}{\sqrt{R^2 + X_L^2}}$$

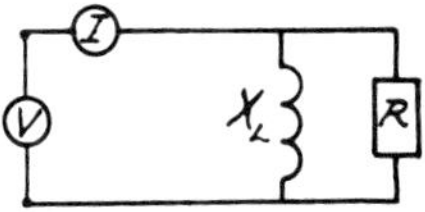

$$Z = \frac{R X_C}{\sqrt{R^2 + X_C^2}}$$

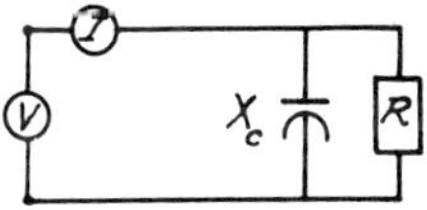

$$Z = \frac{R X_L X_C}{\sqrt{X_L^2 X_C^2 + R^2 (X_L - X_C)^2}}$$

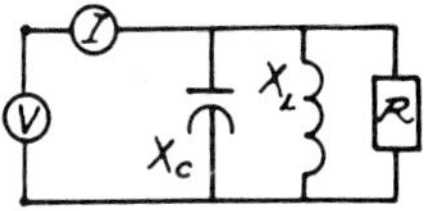

Power Factor

In alternating-current circuits, the product of the voltage and the current is not generally equal to the power as read by a wattmeter. For the special case when the load is purely resistive this product will be equal to power consumed by the circuit.

The power factor is defined as the ratio of the actual power (watts) to the apparent power (volt-amperes) and is equal to the cosine of the phase angle between the impressed voltage and the current in the circuit. The sketch (vector diagram) shown below illustrates how the power factor is related to volt-amperes (or KVA—kilovolt-amperes) and actual watts (or KW—kilowatts). The meaning of reactive volt-amperes, known as vars is also shown by this diagram.

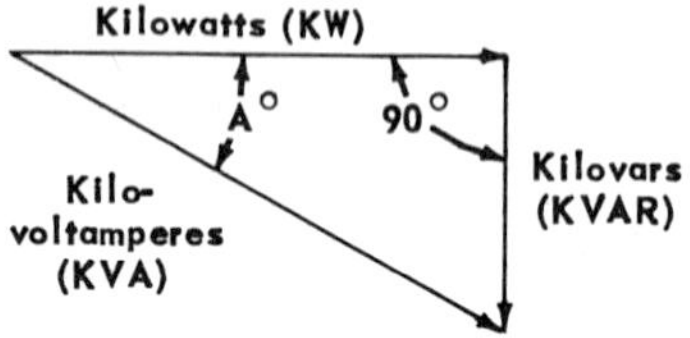

Power Factor	Angle A (Approximately)
1.00	0
.95	18
.90	25
.85	32
.80	36
.75	41
.70	44
.65	50
.60	53
.55	57
.50	60

Since volt-amperes are determined by the current (for a given voltage system), it follows that voltage drop in feeders, etc., as well as machine losses are a function of the kva of a system, whereas the actual or useful power is measured by the horizontal distance shown as kilowatts. Thus, low power factor loads are objectionable. The electric utilities that sell power and energy must transmit a current proportional to the length representing KVA, but the charge (and revenue) for power or energy is determined only by the horizontal distance.

Electric Energy

Electrical energy is measured in kilowatt-hours, which is the product of the power in kilowatts and the time in hours during which this power is used. Kilowatt-hours are measured by a watt-hour meter, which is constructed so that its dial reading—multiplied by several indicated constants—shows the total energy used up to the time of reading. Thus, an average load of 50 kilowatts used for 8 hours a day for 26 days per month, would be an energy load of $50 \times 8 \times 26 = 10{,}400$ kilowatt-hours.

The charge or rate for electrical energy usually consists of three component charges as defined by the so-called Electric Service Classification of the power company. These three items are:

(a) Demand Charge (sliding scale)
(b) Energy Charge (sliding scale)
(c) Fuel Adjustment Charge

The demand charge is usually based on the maximum demand in *kilowatts* during a fixed time interval which may be specified as 5 minutes, 15 minutes, 30 minutes, etc. A common time interval is 30 minutes. This demand is measured by a standard type of demand meter. It records the cumulative or totalized kilowatts (not kilowatt-hours) supplied during such a specified time interval in such a way that each of the component loads is averaged over this time interval. This totalized kilowatt figure is printed on a paper tape at the end of each 30-minute period (usually on the hour and the half-hour).

The energy charge is a sliding scale of charges for the kilowatt-hours of energy consumed (see below).

The fuel adjustment charge accounts for any change in the cost of fuel to the power company used to generate the electrical energy. (see below).

The following service classification is typical of the rate for energy delivered at high voltage to industrial consumers having a minimum demand of 75 kilowatts:

Demand Charge (monthly)

First	50	kilowatts of maximum demand @ $1.50 per kw.
Next	100	kilowatts of maximum demand @ $1.25 per kw.
Excess of	150	kilowatts of maximum demand @ $ .90 per kw.

Energy Charge (monthly)

First	600	kilowatt-hours @ 5¢ per kwhr
Next	4,400	kilowatt-hours @ 4¢ per kwhr
Next	5,000	kilowatt-hours @ 2.2¢ per kwhr

Energy Charge (monthly) (*Cont.*)

Next	260,000 kilowatt-hours @ 1¢ per kwhr	
Excess of	270,000 kilowatt-hours @ 0.5¢ per kwhr	

Fuel Adjustment Charge

The energy charge to the customer for electric energy shall be adjusted upward or downward by 0.017¢ per kilowatt-hour for each 1¢ change in the average cost of fuel above or below 17.5¢ per million BTU.

As an illustration of the above charges for electrical energy, consider the monthly bill for a large industrial consumer:

Demand Charge— 10,560 kw	$ 9,569.00
Energy Charge— 4,923,000 kwhrs	26,181.00
Fuel Adjustment Charge— 4,923,000 kwhrs @ 0.289¢	14,227.47
	$49,977.47

Single Phase Circuits

$$KVA = \frac{VI}{1000}; \quad KW = KVA \times PF$$

Three Phase Circuit—Balanced Star or Wye

$$V_p = \frac{V}{\sqrt{3}} = \frac{V}{1.73} = 0.58\,V;$$

$$V = \sqrt{3}\,V_p = 1.73\,V_p$$

$$I_p = I; \; I_n = 0$$

V = line to line volts

V_p = line to neutral voltage

I = line current

I_p = phase current

I_n = neutral current

Three Phase Circuit—Balanced Delta

$$V = V_p$$

$$I = \sqrt{3}\,I_p = 1.73\,I_p; \quad I_p = \frac{I}{\sqrt{3}} = 0.58\,I$$

Power—Balanced Delta or Wye, 3-Wire, 3-Phase Circuit

$$P = 1.73\ VI \text{ (Power Factor) watts}$$

$$\text{Volt Amperes} = 1.732\ VI$$

$$I = \frac{P}{1.73\ V \text{ (Power Factor)}}$$

Power Loss—D-C or A-C Circuit

$$P = I^2R; \quad I = \sqrt{\frac{P}{R}}; \quad R = \frac{P}{I^2}$$

Three-Phase, Four-Wire Systems

Three-phase, four-wire feeder systems are used extensively in large buildings. This type of system supplies 3-phase, 208 volts for power and single-phase, 120 volts for lighting service from a single bank of transformers. This eliminates the necessity of having two separate systems for 3-phase power and single-phase lighting loads. If used only to supply 120 volts for lighting, it gives the equivalent of three separate circuits.

The permissible *kva* load of a 3-phase, 5-wire system is determined from the formulae for 3-phase in the table above. When this is broken down into three single-phase feeders, the permissible *kva* load in each single-phase feeder is exactly ⅓ that of the total for the 3-phase system.

However, due to the fact that the phase to neutral voltage (usually 120 volts) is only one-third as much as 1.73 times the phase-to-phase voltage (usually 208 volts), the permissible ampere load in each phase is the same as for the entire system.

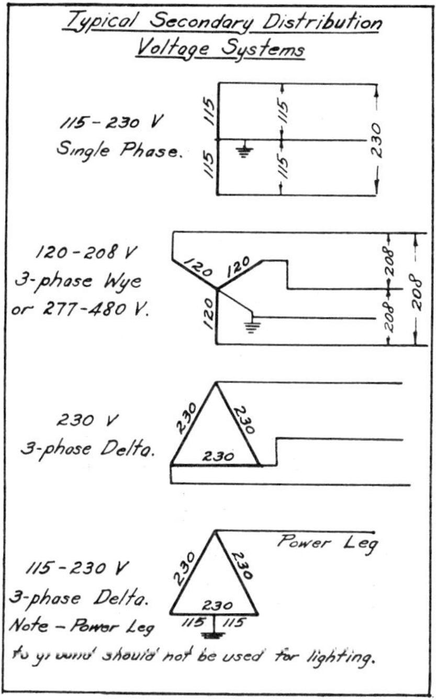

For example, let us assume a 120/208 volt three-phase, 4-wire system having a total *kva* load of 36 *kva*. The current capacity of this system would then be:

$$\frac{kva \times 1000}{1.73 \times \text{voltage}} = \frac{36 \times 1000}{1.73 \times 208} = 100 \text{ amperes}$$

The *kva* load per phase would be:

$$\frac{36}{3} = 12 \; kva \text{ per phase}$$

The current capacity for each single phase would be:

$$\frac{kva \times 1000}{\text{voltage}} = \frac{12 \times 1000}{120} = 100 \text{ amperes}$$

Or, to determine the size of feeders in a 3-phase, 4-wire system:

$$\frac{\text{Total } Kva}{1.73 \times (\text{line to line voltage})} = \frac{\text{Total } Kva}{3 \times (\text{line to neutral voltage})} = \frac{36000}{3 \times 120} = 100 \text{ amperes.}$$

GROUNDING

Good grounding practice should provide (a) safety from electric shock; and (b) protection of equipment from damage by electrical faults.

Safety from Electric Shock

Safety to personnel requires the connection of all non-current carrying metal parts of the electrical system to an equipment ground. Such metal parts would consist of:

1. Metal enclosures, fences, etc., of all apparatus and equipment.
2. Cabinets and conduits.
3. Switchboard instrument cases.
4. Switchboard frames.
5. Instrument transformer cases.
6. Frames of rotating machines; transformer cases, etc.
7. Cable sheaths.
8. Operating handles of disconnect switches; frames, tanks, and operating parts of oil circuit breakers.

In general, any non-current carrying parts with which personnel could come in contact should be at ground or earth potential.

Protection of Equipment

Grounding of the neutral point of the electrical system insures protection of equipment against accidental contact with high voltage resulting from lightning, insulation breakdown in the supply transformer, or other accidental contact with high voltage lines. Such a ground is known as a system ground and is arranged so that protective equipment will operate in case of such faults.

Both the equipment ground and the system ground should be made to the same ground electrode. Ground connections are of two types—those made to water piping systems and those connected to so-called "made grounds," consisting of driven rods, buried cables, or plates. Water piping systems are by far the best, providing that the pipe joints are not made with nonmetallic parts or compounds.

With the present-day use of plastic pipe for water systems, special attention may have to be given to such installations as deep-well pump systems which are to be protected against lightning. In such installations, the pump-motor unit is frequently located at the bottom of the well casing and connected to the water system through a long length of plastic pipe. Since pure water can be a good insulator, it should not be assumed that such a pump-motor unit is well grounded to earth because it is immersed in the water within the metal well casing. As a precaution against unpredictable lightning discharges, the pump-motor assembly should have its frame solidly grounded to the well casing. This would require running a (bare) ground conductor from the pump-motor frame up to a point where this conductor can be fastened to the well casing (or other proper ground bus).

The main requirements for an effective ground bus are low resistance throughout, sufficient current carrying capacity, and permanence. The size of such a ground bus must be large enough so that the ground conductor will not burn off or open up under the largest fault current condition. Table 37 shows minimum recommendations for grounding conductors as they relate to short circuit currents that would flow due to a ground fault. Protection against lightning alone may require a different sizing of such ground conductors.

JOB ESTIMATING

For estimating purposes it should be a basic rule to separate labor cost from material costs and to mark up each of these items according to overhead involved and services rendered.

As in practically all manufacturing and contracting business, the overall cost of supplying and managing labor is much greater than that of merely supplying material. A rough rule of thumb indicates that, on a dollar basis, the overhead cost of supplying labor in the electrical contracting business is approximately $3\frac{1}{2}$ times as great as the overhead cost of supplying material.

As an illustration of the problem, consider a job for which the base cost is $20,000.00, with a 60% material cost and a 40% labor cost. On the basis of the relative proportions of overhead cost mentioned above for labor and material and using an overall mark-up of 20% of the base cost, there results:

	Material	*Labor*
Base Costs	$12,000.00	$ 8,000.00
Mark-Up		
10% on material	1,200.00	
35% on labor		2,800.00
Job Costs	$13,200.00	$10,800.00
Return		
5% on material	660.00	
10% on labor		1,080.00
Totals	$13,860.00	$11,880.00
Price of Job	$25,740.00	

The total mark-up amounts to $4000.00, being 20% of the base cost, with the mark-up on labor costs being taken at $3\frac{1}{2}$ times (35%) of the mark-up on material (10%). The difference between the price of the job and the base cost is $5,740.00, of which the labor item is $3,880.00.

This illustration emphasizes the importance of estimating labor costs as closely as possible. Clearly, it is a straightforward job to prepare a list of materials for an installation and get a reliable cost figure for all the items. On the other hand, a reliable estimate of the base labor cost, or labor time at a known hourly rate, is a much more difficult and much more important problem, since the chances for error are far greater.

Present-day practice for finding the labor cost item of a job estimate makes use of so-called "Labor Units." These are units of man-hours required for each type of job and represent an average or composite figure for favorable working conditions, skilled mechanics, and large-scale installations. The most complete data in this field has been assembled by the National Electrical Contractors Association (NECA) and is available to members of the Association.

It should be emphasized that these labor-unit figures can be completely misleading and meaningless under many different working conditions which are likely to exist. In the hands of an experienced estimator such tables are almost a necessity; in the hands of a novice, such tables may do more harm than good. For example, the NECA Manual of Labor Units lists all labor cost data on a net man-hour basis and then applies a so-called "Job Factor," which may increase the original man-hour figure by as much as 50%, depending on such factors as the type of building, working conditions, attitude of the general contractor, non-productive labor, etc.

Another important factor in estimating labor time is the weight of equipment involved and the height at which work is being done. Slow speed motors are heavier than high speed motors of the same horsepower; oil-cooled transformers are heavier than air-cooled units of the same rating, especially in the smaller sizes; 25-cycle equipment is much heavier than 60-cycle. The weights of cabinets and pull boxes can be an important item in their installation. Tables 33, 34, 35, 36 givc thc wcights of typical motors, transformers, and pull boxes. Table 32(a) gives the weights of conduit.

With these reservations and warnings, some typical man-hour figures for certain complete installations are tabulated in the following pages. They are subject to wide variations depending on working conditions, use of labor-saving devices, new or existing construction, and many other unforeseen situations. Perhaps one of the most important items in such estimates of fixture work and conduit runs is that of ceiling or mounting height.

Raceway and Fittings—12 ft. ceiling—single run	*Hours per 100 ft.*
Conduit—Straight	
1½"	8
2"	11
3"	18.5
4"	30
Conduit—Elbows	
1½"	75
2"	120
3"	300
4"	400
L-Type Condulets	*Hours each*
1½"	2.0
2"	2.5
3"	4.8
4"	7.3
Hangers and Fastenings—12 ft. ceiling	*Hours per 100 ft.*
¼" Tamp-in anchors	19
¼" Toggle bolts	15
¼" Machine bolts in steel flange	23
Beam clamps	15
Split Pipe Rings and Sockets	
1½"	18
2"	23
3"	33
4"	48
One-Hole Straps	
1½"	2
2"	3
3"	5
4"	8

Pipe-Rise Clamps

2"	40
3"	47
4"	60

Service and Feeder Wiring

Single conductor, 4-wire, 100-ft run, average conditions, type R, RH stranded.

Number or Size	*Hours per 1000 ft*
6	17
4	20
2	22
0	31
00	35
0000	52
250 MCM	55
500 MCM	70
1000 MCM	120

Solderless Connectors—Hours Each

6	0.6
2	0.8
0	1.0
0000	1.5
250 MCM	1.6
500 MCM	2.4
1000 MCM	4.2

Bus Duct

12-foot ceiling, concrete, 3-pole, feeder type.

600-ampere duct, hours per 100 ft.	50
600-ampere duct, hanger, hours each	1.5
600-ampere duct, elbow, hours each	6
600-ampere duct, feeder tap box, hours each	18

BRANCH CIRCUIT WIRING

12-ft ceilings, concealed, favorable working conditions.

	Wood Frame	Flat Slab
Conduit—rigid		
½"—hrs/100 ft	5.5	3.0
¾"—hrs/100 ft	5.8	3.8
1" —hrs/100 ft	7.5	5.5
EM Tubing		
½"—hrs/100 ft	5.0	3.0
¾"—hrs/100 ft	5.4	3.5
1" —hrs/100 ft	7.0	5.0
Armored Cable or Non-Metallic Sheathed Cable		
2-wire #12, hrs/1000 ft	4.0	
3-wire #12, hrs/1000 ft	4.5	
2-wire # 8, hrs/1000 ft	6.5	
3-wire # 8, hrs/1000 ft	7.5	
Branch Circuit Wire-Solid		
#12, hrs/1000 ft	7.5	7.5
#10, hrs/1000 ft	11.0	11.0
# 8, hrs/1000 ft	13.0	13.0
Outlet Boxes—hrs each		
4" Octagon box, ½" K.O.	0.6	0.5
4" Box with round cover ball aligner	0.9	0.7
3-Gang switch box	0.7	0.8
2-Gang floor box		2.5
Switches and Receptacles complete with plates—hrs each		
20-amp double-pole switch	0.75	
10-amp three-way ceiling pull switch	0.75	
20-amp three-wire receptacle	0.5	

Commercial Fixtures—
Assembled and hung,
12-ft ceiling, hrs each.

Ceiling unit with glassware, 300-500 watts 0.5
Stem suspension unit, 300-500 watts 0.75
Flush units 2.0

Industrial Units—(as above)

RLM Dome, suspended, 300-500 watts 0.75
Glassteel reflector with glassware (ditto) 0.8

Fluorescent Fixtures

Subject to wide variations, ceiling or pendant, 12-ft ceiling, hrs each.

4-ft—4 lamp, open type, all metal	2.5
6-ft—4 lamp, open type, all metal	3.0
8-ft—4 lamp, open type, all metal	3.3
8-ft—8 lamp, open type, all metal	4.5
High-intensity, mercury-vapor unit, supporting bracket, fixture, transformer, 20-ft height	5.0

Surface Metal Raceway—
12-ft ceiling, hrs each.

	Wood	Concrete
Raceway per 100 ft	8	14
Ceiling boxes	0.8	1.0
Wall boxes	0.6	0.8
Corner boxes	0.5	0.7
Flat elbows	0.2	0.3
Tees	0.5	0.6
Wire—#12/1000 ft	12	12

GRAPHICAL SYMBOLS FOR ELECTRICAL DIAGRAMS

(American Standard)

Disconnecting Switch, Single Throw

Disconnecting Switch, Gang Operated

Disc. Sw., Single Blade, Double Throw

Automatic Throw-Over

Disconnecting Fuse

Reactor (Non-Magnetic Core)

Reactor (Magnetic Core)

Resistor

Static Condenser

Valve Type Lightning Arrester

Rheostat, Hand Operated

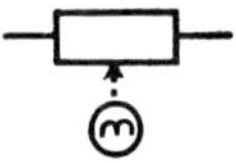

Rheostat, Motor Operated

Shunt

Fuse

Battery

Ground

Pothead, Non-Disconnecting

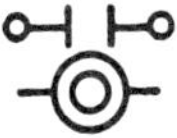

Magnetic Contactor

Half Wave Rectifier

Full Wave Rectifier

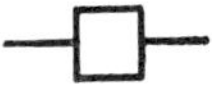

Oil Circuit Breaker, Automatic, Single Throw

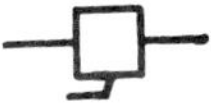

Oil Circuit Breaker, Double Throw

Oil Circuit Breaker, Removable Type

Air Circuit Breaker, Automatic

Air Circuit Breaker, Motor Operated

Air Circuit Breaker,
Solenoid Operated

Drawout Air Circuit
Breaker

Network Protector

Squirrel Cage Motor

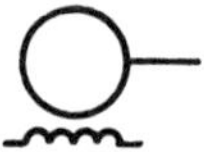

Synchronous Motor

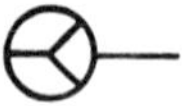

3-Phase "Wye"
Connection

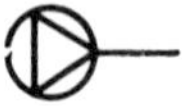

3-Phase "Delta"
Connection

Wound Rotor Induction Motor

D.C. Motor with
Shunt Field

D.C. Motor with
Series Field

1-Phase Power
Transformer

3-Phase Power Transformer, Connected "Delta-Delta"

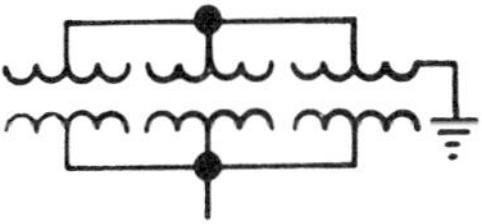

3—Single Phase Trans. in 3-Ph. Bank, Grounded Neutral

Potential Transformer

Current Transformer

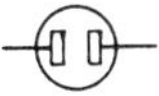

2-conductor non-polarized connector with female contacts

2-conductor non-polarized connector with male contacts

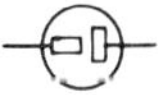

2-conductor polarized connector with female contacts

2-conductor polarized connector with male contacts

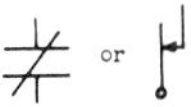

Closed contact (break)

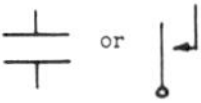

Open contact (make)

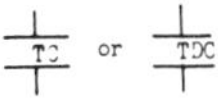

Open contact with time closing feature (TC or TDC)

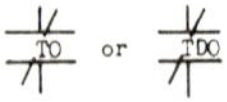

Closed contact with time opening feature (TO or TDO)

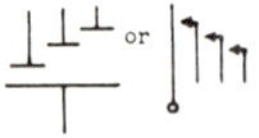

Time sequential closing

METER OR INSTRUMENT DESIGNATION

Letter combination within a circle

A	Ammeter
AH	Ampere-hour meter
CMA	Contact-making (or breaking) ammeter
CMC	Contact-making (or breaking) clock
CMV	Contact-making (or breaking) voltmeter
CRO	Oscilloscope or cathode-ray oscillograph
D	Demand meter
DB	DB (decibel) meter
DBM	DBM (decibels referred to 1 milliwatt) meter
DTR	Demand-totalizing relay
F	Frequency meter
G	Galvanometer
GD	Ground detector
I	Indicating
M	Integrating
μA or UA	Microammeter
MA	Milliammeter
N	Noise meter
OHM	Ohmmeter
OP	Oil pressure
OSCG	Oscillograph, string
PH	Phase meter
PI	Position indicator
PF	Power-factor meter

RD	Recording demand meter
REC	Recording
RF	Reactive-factor meter
S	Synchroscope
TLM	Telemeter
T	Temperature meter
TT	Total time
VH	Varhour meter
V	Voltmeter
VA	Volt-ammeter
VAR	Varmeter
VI	Volume indicator
VU	Standard volume indicator
W	Wattmeter
WH	Watthour meter

GRAPHICAL ELECTRICAL SYMBOLS FOR ARCHITECTURAL PLANS

(American Standards Association)

Ceiling	Wall	General Outlets
○	-○	Outlet
Ⓑ	-Ⓑ	Blanked Outlet
Ⓓ		Drop Cord
Ⓔ	-Ⓔ	Electric Outlet; for use only when circle used alone might be confused with columns, plumbing symbols, etc
Ⓕ	-Ⓕ	Fan Outlet
Ⓙ	-Ⓙ	Junction Box
Ⓛ	-Ⓛ	Lamp Holder
Ⓛ$_{PS}$	-Ⓛ$_{PS}$	Lamp Holder with Pull Switch
Ⓢ	-Ⓢ	Pull Switch

Ⓥ	-Ⓥ	Outlet for Vapor Discharge Lamp
Ⓧ	-Ⓧ	Exit Light Outlet
©	-©	Clock Outlet (Specify Voltage)

Convenience Outlets

	Duplex Convenience Outlet
	Convenience Outlet other than Duplex 1 = Single, 3 = Triplex, etc
	Weatherproof Convenience Outlet
	Range Outlet
	Switch and Convenience Outlet
	Radio and Convenience Outlet
	Special Purpose Outlet (Des. in Spec.)
	Floor Outlet

Switch Outlets

S	Single Pole Switch
S_2	Double Pole Switch
S_3	Three Way Switch
S_4	Four Way Switch
S_D	Automatic Door Switch
S_E	Electrolier Switch
S_K	Key Operated Switch
S_P	Switch and Pilot Lamp
S_{CB}	Circuit Breaker

S_{WCB}	Weatherproof Circuit Breaker
S_{MC}	Momentary Contact Switch
S_{RC}	Remote Control Switch
S_{WP}	Weatherproof Switch
S_F	Fused Switch
S_{WF}	Weatherproof Fused Switch

Special Outlets

$\bigcirc_{a,b,c,etc}$ $\ominus_{a,b,c,etc}$ $S_{a,b,c,etc}$	Any standard symbol as given above with the addition of a lower case subscript letter may be used to designate some special variation of standard equipment of particular interest in a specific set of architectural plans. When used they must be listed in the Key of Symbols on each drawing and if necessary further described in the specifications.

Panels, Circuits, and Miscellaneous

	Lighting Panel
	Power Panel
——	Branch Circuit; Concealed in Ceiling or Wall
— —	Branch Circuit; Concealed in Floor
- - - -	Branch Circuit; Exposed

Home Run to Panel Board. Indicate number of circuits by number of arrows.
Note: Any circuit without further designation indicates a two-wire circuit. For a greater number of wires indicate as follows: —///— (3 wires) —////— (4 wires), etc.

Feeders. Note: Use heavy lines and designate by number corresponding to listing in Feeder Schedule.

Underfloor Duct and Junction Box. Triple System. Note: For double or single systems eliminate one or two lines. This symbol is equally adaptable to auxiliary system layouts.

Generator

Motor

Instrument

Power Transformer (Or draw to scale.)

Controller

Isolating Switch

Auxiliary Systems

Pushbutton

Buzzer

	Bell
	Annunciator
	Outside Telephone
	Interconnecting Telephone
	Telephone Switchboard
	Bell-Ringing Transformer
	Electric Door Opener
	Fire Alarm Bell
	Fire Alarm Station
	City Fire Alarm Station
	Fire Alarm Central Station
	Automatic Fire Alarm Device
	Watchman's Station
	Watchman's Central Station
	Horn
	Nurse's Signal Plug
	Maid's Signal Plug
	Radio Outlet
	Signal Central Station
	Interconnection Box
	Battery
	Auxiliary System Circuits

Note: Any line without further designation indicates a 2-wire system. For a greater number of wires designate with numerals in manner similar to — - — 12-No. 18W-

3/4″C., or designate by number corresponding to listing in Schedule.

$\square_{a,b,c}$ Special Auxiliary Outlets. Subscript letters refer to notes on plans or detailed description in specifications.

USEFUL MOTOR APPLICATION DATA

Horsepower Required

Pumps

$$\text{Water Hp.} = \frac{\text{GPM} \times \text{IDH (in feet)}}{3960}$$

$$= \frac{\text{GPM} \times \text{IDH (in lbs./sq.in.)}}{1715}$$

$$\text{Motor Hp.} = \frac{\text{Water Hp.}}{\text{Pump Efficiency}}$$

$$\text{KW Input} = \frac{\text{GPM} \times \text{IDH (in feet)}}{5308 \times \text{pump eff.} \times \text{motor eff.}}$$

GPM = gallons per minute of water
IDH = total dynamic head in feet of water.

The above formulae are for cold water of density (specific gravity) equal to 1. The horsepower for liquids of different densities is obtained by multiplying the above water horsepower by the density of the liquid.

1 lb./sq.in. = 2.31 ft. of water
1 lb./sq.in. = 2.04 ins. of mercury

1 ft. of water	= 0.433 lb./sq.in.
1 gallon	= 231 cu. ins. = 3.785 liters
1 cu. ft.	= 7.48 gallons
1 U.S. gallon	= 8.33 pounds
1 cu. ft. water	= 62.4 pounds
1 cu. ft./sec.	= 448.8 GPM

Fans and Blowers

Motor Hp. = H

$$\frac{\text{CFM} \times \text{pressure in ins. of water}}{6356 \times \text{Fan Efficiency}}$$

An average ventilating fan has a typical efficiency of about 65 per cent.

Elevators

$$\text{Approx. Hp.} = \frac{(L - W)\ S}{33{,}000 \times \text{Efficiency}}$$

L = rated elevator load in pounds without counterweight.
W = weight in excess of that required to counterbalance empty car.
S = lift speed in feet per minute.

Escalators

For the usual stairway speeds, a 3-foot stairway with a rise of 17 ft. requires about 12 hp. when fully loaded.

Hoists

$$\text{Approx. Hp.} = \frac{W \times S \times \text{sine } A}{33{,}000 \times \text{Efficiency}}$$

W = weight in pounds
S = feet per minute
A = angle of hoist with horizontal

Air Conditioning

Multiply the number of tons of refrigeration required by:

1.25 Hp per ton for ½ to 5 tons capacity.
1.1 Hp per ton for 5 to 50 tons capacity.
1.0 Hp per ton for capacity above 50 tons.

Tons of air conditioning required to remove a lighting system heat load are equal to

$$\frac{\text{Lighting Kilowatts} \times 3414}{12000}$$

(Equivalent to: one ton of air conditioning capacity will remove the heat for a lighting load of 3500 watts).

Ice Making, Main Compressor

Multiply the number of tons of ice required per 24 hour day by:

3.5 Hp per ton for capacity up to 50 tons capacity.
3.25 Hp per ton for 50 60 200 tons capacity.
3.1 Hp per ton for capacity over 200 tons.

Formulae for Determining Amperes, Hp., Kw. and Kva

To Find	Direct Current	Alternating Current	
		Single Phase	Three Phase
Amperes when Horsepower is known	$\frac{Hp. \times 746}{V \times \%\ Eff.}$	$\frac{Hp. \times 746}{V \times \%\ Eff. \times P.F.}$	$\frac{Hp. \times 746}{1.73 \times V \times \%\ Eff. \times P.F.}$
Amperes when Kilowatts is known	$\frac{Kw \times 1000}{V}$	$\frac{Kw \times 1000}{V \times P.F.}$	$\frac{Kw \times 1000}{1.73 \times V \times P.F.}$
Amperes when Kva is known		$\frac{Kva \times 1000}{V}$	$\frac{Kva \times 1000}{1.73 \times V}$
Kilowatts	$\frac{I \times V}{1000}$	$\frac{I \times V \times P.F.}{1000}$	$\frac{I \times V \times 1.73 \times P.F.}{1000}$
Kva		$\frac{I \times V}{1000}$	$\frac{I \times V \times 1.73}{1000}$
Horsepower (output)	$\frac{I \times V \times \%\ Eff.}{746}$	$\frac{I \times V \times \%\ Eff. \times P.F.}{746}$	$\frac{I \times V \times 1.73 \times \%\ Eff. \times P.F.}{746}$

V = volts; I = amperes; P.F. = power factor; Hp = horsepower; Kw = kilowatts; Kva. = kilovolt amperes; % Eff. = per cent efficiency.

Approximate Motor Amperes per Terminal:
230 V. d-c—4 amperes per Hp.; 3-phase, 220 V. a-c—2.5 amperes per Hp.; 3-phase, 440 V. a-c—1.25 amperes per Hp.; 3-phase, 550 V. a-c—1 ampere per Hp.

EXAMPLES

The following examples show typical methods of calculating and selecting conductor sizes for representative installations. Refer to Tables 2 (310-12), 3 (310-14), 4 (310-13), 5 (310-15) and the Notes pertaining to such tables. For the computation of range loads in these examples Column A of Table 25 (220-5) has been used. For optional methods, see Columns B and C of that Table. All Section and Table references are to the 1971 NEC.

Example No. 1. Single-Family Dwelling

Dwelling has a floor area of 1,500 sq. ft. exclusive of unoccupied cellar. unfinished attic, and open porches. It has a 12-kw range.

Computed Load (see Section 220-4)

General Lighting Load:

1,500 sq. ft. at 3 watts per sq. ft. = 4,500 watts.

Minimum Number of Branch Circuits Required (see Section 220-3)

General Lighting Load:

4,500 ÷ 115 = 39.1 amperes: or three 15-ampere 2-wire circuits; or two 20-ampere 2-wire circuits

Small Appliance Load: Two 2-wire 20-ampere circuits (Section 220-3(b))

Laundry Load: One 2-wire 20-ampere circuit (Section 220-3(b))

Minimum Size Feeders Required (see Section 220-4)

Computed Load

General Lighting	4,500 watts
Small Appl. Load	3,000 watts
Laundry	1,500 watts
Total (without range)	9,000 watts
3,000 watts at 100%	3,000 watts
9,000 − 3,000 = 6,000 watts at 35% = ...	2,100 watts
Net computed load (without range).	5,100 watts
Range Load (see Table 220-5).	8,000 watts
Net computed load (with range).	13,100 watts

For 115/230-volt 3-wire system feeders; 13,100 ÷ 230 = 57 amperes

Net computed load exceeds 10 kw. so service conductors shall be 100 amperes (see Section 230-41, Exception No. 1).

Example No. 1(a). Single-Family Dwelling

Same conditions as Example No. 1, plus addition of one 6-ampere 230-volt room air-conditioning unit and three 12-ampere 115-volt room air-conditioning units. See Article 422, Part F.

From Example No. 1, feeder current is 57 amperes (3 wire, 230 volt)

Line A	Neutral	Line B	
57		57	amperes from Example No. 1
6		6	one 230-volt air cond. motor
12		12	two 115-volt air cond. motors
—		12	one 115-volt air cond. motor
3		3	25% of largest motor (Section 430-24)
78		90	amperes per line

Example No. 1(b). Single-Family Dwelling
Optional Calculation for Single-Family Dwelling (Section 220-7)

Dwelling has a floor area of 1,500 sq. ft. exclusive of unoccupied cellar, unfinished attic and open porches. It has a 12-kw range, a 2.5-kw water heater, a 1.2-kw dishwasher, 9 kw of electric space heating installed in five rooms, a 5-kw clothes dryer, and a 6-amp. 230-volt room air-conditioning unit.

Air conditioner kw is 6 × 230 ÷ 1000 = 1.38 kw

1.38 kw is less than the connected load of 9 kw of space heating; therefore, the air conditioner load need not be included in the service calculation (see Section 220-4(I)).

1,500 sq. ft. at 3 watts	4.5 kw
Two 20-amp. appliance outlet circuits at 1,500 watts each	3.0 kw
Laundry circuit	1.5 kw
Range (at nameplate rating)	12.0 kw
Water heater	2.5 kw
Dishwasher	1.2 kw
Space heating	9.0 kw
Clothes dryer	5.0 kw
	38.7 kw

First 10 kw at 100% = 10.00 kw

Remainder at 40% (28.7 kw × .4) = 11.48 kw

Calculated load for service size 21.48 kw = 21,480 watts

21,400 ÷ 230 = 93 amperes

Therefore, this dwelling may be served by a 100-ampere service.

Example No. 1(c). Single-Family Dwelling
Optional Calculation for Single-Family Dwelling (Section 220-7)

Dwelling has a floor area of 1,500 sq. ft. exclusive of unoccupied cellar, unfinished attic and open porches. It has two 20-ampere small appliance circuits, one 20-ampere laundry circuit, two 4-kw wall-mounted ovens, one 5.1-kw counter-mounted cooking unit, a 4.5-kw water heater, a 1.2-kw dishwasher, a 5-kw combination clothes washer and dryer, six 7-ampere 230-volt room air-conditioning units and a 1.5-kw permanently installed bathroom space heater.

Air conditioning kw calculation

Total amperes 6 × 7 = 42.00 amperes
25% of largest motor .25 × 7 = 1.75 amperes
43.75 amperes

43.75 × 230 ÷ 1,000 = 10.1 kw of air-conditioner load

Load included at 100%

Air conditioning . . . . 10.1 kw
Space heater (omit, see Section 220-4(I))

Other Load	kW
1,500 sq. ft. at 3 watts.	4.5
Two 20-amp small appliance circuits at 1,500 watts	3.0
Laundry circuit.	1.5
2 ovens	8.0
1 cooking unit	5.1
Water heater	4.5
Dishwasher	1.2
Washer/dryer	5.0
Total other load.	32.8

1st 10 kw at 100% . . . . 10.0 kw
Remainder at 40% (22.8 kw X .4). 9.12 kw
Total calculated load . . . . 29.22 kw = 29,220 watts

29,220 ÷ 230 = 127 amperes (service rating)

Example No. 2. Small Roadside Fruitstand with No Show Windows

A small roadside fruitstand with no show windows has a floor area of 150 square feet. The electrical load consists of general lighting and a 1,000-watt floodlight. There are no other outlets.

Computed Load (Section 220-4)

*General Lighting
150 sq. ft. at 3 watts/sq. ft. X 1.25 = 562 watts
(3 watts/sq. ft. for stores)
562 watts ÷ 115 = 4.88 amperes

One 15-ampere 2-wire branch circuit required (Section 220-3)

Minimum Size Service Conductor Required (Section 230-41 Exception No. 2)

Computed load.	562 watts
Floodlight load.	1,000 watts
Total load	1,562 watts

1,562 ÷ 115 = 13.6 amperes

Use No. 8 service conductor (Section 230-41, Exception No. 2)

Use a 30-ampere service switch or breaker (Section 230-71(a), Exception No. 2)

Example No. 3. Store Building

A store 50 feet by 60 feet, or 3,000 square feet, has 30 feet of show window.

Computed Load (Section 220-4)

*General lighting load:
3,000 square feet at 3 watts per square foot X 1.25. . . . . 11,250 watts

*The above examples assume that the entire general lighting is a continuous load and the load is therefore increased by 25 percent in accordance with Section 220-2. The 25 percent increase is not applicable to any portion of the load that is not continuous.

**Show window lighting load:

30 feet at 200 watts per foot . . . . 6,000 watts

Minimum Number of Branch Circuits Required (Section 220-3)

***General lighting load: 11,250 ÷ 230 = 49 amperes for 3 wire, 115/230 volts; or 98 amperes for 2 wire, 115 volts:

Three 30-ampere, 2-wire; and one 15-ampere, 2-wire circuits; or
Five 20-ampere, 2-wire circuits; or
Three 20-ampere, 2-wire, and three 15-ampere, 2-wire circuits; or
Seven 15-ampere, 2-wire circuits; or
Three 15-ampere, 3-wire, and one 15-ampere, 2-wire circuits.

Special lighting load (show window): (Sections 220-2 Exception No. 1 and 220-4(c)): 6,000 ÷ 230 = 26 amperes for 3-wire, 115/230 volts; or 52 amperes for 2-wire, 115 volts:

Four 15-ampere, 2-wire circuits; or
Three 20-ampere, 2-wire circuits; or
Two 15-ampere, 3-wire circuits.

Minimum Size Feeders (or Service Conductors) Required (Section 215-2):

For 115/230-volt, 3-wire system:

Ampere load: 49 plus 26 = 75 amperes. (Section 220-2).

For 115-volt system:

Ampere load: 98 plus 52 = 150 amperes. (Section 220-2).

**If show window load is computed as per Section 220-2, the unit load per outlet shall be increased 25 percent.

***The load on each general lighting branch circuit shall not exceed 80 percent of the branch-circuit rating (Section 210-23b).

Example No. 4. Multi-Family Dwelling

Multi-family dwelling having a total floor area of 32,000 square feet with 40 apartments.

Meters in two banks of 20 each and individual sub-feeders to each apartment.

One-half of the apartments are equipped with electric ranges of not exceeding 12 kw each.

Area of each apartment is 800 square feet.

Laundry facilities on premises available to all tenants. Add no circuit to individual apartment. Add 1,500 watts for each laundry circuit to house load and add to the example as a "house load."

Computed Load for Each Apartment (Article 220):

General lighting load:

800 square feet at 3 watts per square foot. . . . . 2,400 watts

Special appliance load:

Electric range . . . . 8,000 watts

Minimum Number of Branch Circuits Required for Each Apartment (Section 220-3):

General lighting load: 2,400 ÷ 115 = 21 amperes or two 15-ampere, 2-wire circuits; or two 20-ampere, 2-wire circuits.

Small appliance load: Two 2-wire circuits of No. 12 wire. (See Section 220-3(b)).

Range circuit: 8,000 ÷ 230 = 35 amperes or a circuit of two No. 8's and one No. 10 as permitted by Section 210-19(c).

Minimum Size Sub-Feeder Required for Each Apartment (Section 215-2):

Computed load (Article 220):	
General lighting load	2,400 watts
Small appliance load, two 20-ampere circuits . .	3,000 watts
Total computed load (without ranges)	5,400 watts
Application of Demand Factor:	
3,000 watts at 100%	3,000 watts
2,400 watts at 35%	840 watts
Net computed load (without ranges)	3,840 watts
Range load.	8,000 watts
Net computed load (with ranges)	11,840 watts

For 115/230-volt, 3-wire system (without ranges):
Net computed load, 3,840 ÷ 230 = 16.7 amperes.
Size of each sub-feeder (see Section 215-2).

For 115/230-volt, 3-wire system (with ranges):
Net computed load, 11,840 ÷ 230 = 51.5 amperes.

Sub-Feeder Neutral:	
Lighting and small appliance load	3,840 watts
Range load, 8,000 watts at 70% (see Section 220-4(e))	5,600 watts
Net computed load (neutral)	9,440 watts

9,440 ÷ 230 = 41 amperes

Minimum Size Feeders Required from Service Equipment to Meter Bank
(For 20 Apartments–10 with Ranges):

Total Computed Load:	
Lighting and small appliance load, 20 × 5,400 .	108,000 watts
Application of Demand Factor:	
3,000 watts at 100%	3,000 watts
105,000 watts at 35%	36,750 watts
Net computed lighting and small appliance load	39,750 watts
Range Load, 10 ranges (less than 12 kw; Col. A, Table 220-5)	25,000 watts
Net computed load (with ranges)	64,750 watts

For 115/230-volt, 3-wire system:
Net computed load, 64,750 ÷ 230 = 282 amperes.

Feeder Neutral:	
Lighting and small appliance load	39,750 watts
Range load: 25,000 watts at 70% (see Section 220-4(e))	17,500 watts
Computed load (neutral)	57,250 watts

57,250 ÷ 230 = 249 amperes.

Further Demand Factor (Section 220-4(e)):	
200 amperes at 100%	= 200 amperes
49 amperes at 70%	= 34 amperes
Net computed load (neutral)	234 amperes

Minimum Size Main Feeder (or Service Conductors) Required
(For 40 Apartments–20 with Ranges):

Total Computed Loads:	
Lighting and small appliance load, 40 × 5,400 .	216,000 watts

Application of Demand Factor:

3,000 watts at 100%	3,000 watts
117,000 watts at 35%	40,950 watts
96,000 watts at 25%	24,000 watts
Net computed lighting and small appliance load	67,950 watts
Range load, 20 ranges (less than 12 kw, Col. A, Table 220-5)	35,000 watts
Net computed load	102,950 watts

For 115/230-volt, 3-wire system:

Net computed load, 102,950 ÷ 230 = 448 amperes.

Feeder Neutral:

Lighting and small appliance load	67,950 watts
Range load, 35,000 watts at 70% (see Section 220-4(e))	24,500 watts
Computed load (neutral)	92,450 watts

92,450 ÷ 230 = 402 amperes.

Further Demand Factor (see Section 220-4(e)):

200 amperes at 100%	= 200 amperes
202 amperes at 70%	= 141 amperes
Net computed load (neutral)	341 amperes

See Tables 310-12 through 310-15, Notes 8 and 10.

Example No. 4(a). Optional Calculation for Multi-Family Dwelling

Multi-family dwelling equipped with electric cooking and space heating or air conditioning and having a total floor area of 32,000 square feet with 40 apartments.

Meters in two banks of 20 each plus house metering and individual sub-feeders to each apartment.

Each apartment is equipped with an electric range of 8-kW nameplate rating, four 1.5-kW separately controlled 230-volt electric space heaters, and a 2.5-kW 230-volt electric water heater.

A common laundry facility is available to all tenants (Section 210-22(b), Exception No. 1).

Area of each apartment is 800 square feet.

Computed Load for Each Apartment (Article 220)

General Lighting Load:

800 square ft. at 3 watts per sq. ft.	2,400 watts
Electric range	8,000 watts
Electric heat 6 kw (or air conditioning if larger)	6,000 watts
Electric water heater	2,500 watts

Minimum Number of Branch Circuits Required for Each Apartment

General lighting load 2,400 watts ÷ 115 = 21 amperes or two 15-ampere 2-wire circuits or two 20-amp 2-wire circuits.

Small appliance load: Two 2-wire circuits of No. 12 (See Section 220-3(b))

Range circuit 8,000 watts × 80% ÷ 230 = 28 amperes on a circuit of three No. 10 as permitted in Column C of Table 220-5

Space Heating 6,000 watts ÷ 230 = 26 amperes

No. of circuits (See Section 220-3(a))

Minimum Size Sub-Feeder Required for Each Apartment (Section 215-2)

Computed load (Article 220):

General lighting load	2,400 watts
Small appliance load, two 20-amp circuits	3,000 watts
Total computed load (without range and space heating	5,400 watts

Application of Demand Factor:

3,000 watts at 100%	3,000 watts
2,400 watts at 35%	840 watts
Net computed load (without range and space heating)	3,840 watts
Range load	6,400 watts
Space heating Section 220-4(f)	6,000 watts
Water heater	2,500 watts
Net computed load for individual apartment	18,740 watts

For 115/230-volt 3-wire system:
Net computed load 18,740 ÷ 230 = 81 amperes

Sub-Feeder Neutral (Section 220-4(e)):

Lighting and small appliance load	3,840 watts
Range load 6,400 watts at 70% (see Section 220-4(e))	4,480 watts
Space and water heating (no neutral) 230 volt	0 watts
Net computed load (neutral)	8,320 watts

8,320 ÷ 230 = 36 amperes

Minimum Size Feeder Required from Service Equipment to Meter Bank for 20 Apartments

Total Computed Load:

Lighting and small appliance load 20 X 5,400	108,000 watts
Water and space heating load 20 X 8,500	170,000 watts
Range load 20 X 8,000	160,000 watts
Net computed load (20 apartments)	438,000 watts

Net Computed using Optional Calculation (Table 220-9)

438,000 X .38	166,440 watts

166,440 ÷ 230 = 724 amperes

Feeder Neutral:

Lighting and small appliance load 108,000 X 38% (Table 220-9)	41,040 watts
Range load 8,000 X 20 = 160,000 watts 160,000 X 38% (Table 220-9)	60,800 watts
computed load (neutral)	101,840 watts

101,840 ÷ 230 = 443 amperes

Minimum Size Mains Feeder Required (less house load)
(For 40 Apartments)

Total Computed Load:

Lighting and small appliance load 40 X 5,400	216,000 watts
Water and space heating 40 X 8,500	340,000 watts
Range load 40 X 8,000	320,000 watts
Net computed load (40 apartments)	876,000 watts
Net computed using Optional Calculation (Table 220-9) 876,000 X 28%	245,280 watts

245,280 ÷ 230 = 1,066 amperes

Feeder Neutral:

Lighting and small appliance load 40 × 5,400 × 28% (Table 220-9). . . . . 60,480 watts
Range load 8,000 × 40 × 28% (Table 220-9) . . 89,600 watts
Computed load (neutral). . . . . .150,080 watts
150,080 ÷ 230 = 653 amperes

Add to obtain size of service conductors, the entire house load including laundry circuit(s) in accordance with applicable Sections of Article 220.

Example No. 5. Calculation of Feeder Neutral (See Section 220-4)

The following example illustrates the method of calculating size of a feeder neutral for the computed load of a 5-wire, 2-phase system, where it is desired to modify the load in accordance with provisions of Section 220-4.

An installation consisting of a computed load of 250 amperes connected between the feeder neutral and each ungrounded feeder conductor.

Feeder Neutral (maximum unbalance of load 250 amp. × 140% = 350 amperes):

200 amperes (first)	at 100% =	200 amperes
150 amperes (excess)	at 70% =	105 amperes
Computed load		305 amperes

Example No. 6. Maximum Demand for Range Loads

Table 220-5, Column A applies to ranges not over 12 kw. The application of Note 1 to ranges over 12 kw (and not over 27 kw) is illustrated in the following examples:

A. *Ranges all of same rating.*

Assume 24 ranges each rated 16 kw.
From Column A the maximum demand for 24 ranges of 12 kw rating is 39 kw.
16 kw exceeds 12 kw by 4.
5% × 4 = 20% (5% increase for each kw in excess of 12).
39 kw × 20% = 7.8 kw increase.
39 + 7.8 = 46.8 kw: value to be used in selection of feeders.

B. *Ranges of unequal rating.*

Assume 5 ranges each rated 11 kw.
2 ranges each rated 12 kw.
20 ranges each rated 13.5 kw.
3 ranges each rated 18 kw.

5 × 12	=	60	Use 12 kw for range rated less than 12.
2 × 12	=	24	
20 × 13.5	=	270	
3 × 18	=	54	
30		408 kw	

408 ÷ 30 = 13.6 kw (average to be used for computation)

From Column A the demand for 30 ranges of 12 kw rating is 15 + 30 = 45 kw.

13.6 exceeds 12 by 1.6 (use 2.).
5% × 2 = 10% (5% increase for each kw in excess of 12).
45 kw × 10% = 4.5 kw increase.
45 + 4.5 = 49.5 kw = value to be used in selection of feeders.

Example No. 7. Ranges on a 3-Phase System (Section 220-4(j))

Thirty ranges rated at 12 kw each are supplied by a 3-phase, 4-wire, 120/208-volt feeder, 10 ranges on each phase.

As there are 20 ranges connected to each ungrounded conductor, the load should be calculated on the basis of 20 ranges (or in case of unbalance, twice the maximum number between any two phase wires) since diversity applies only to the number of ranges connected to adjacent phases and not the total.

The current in any one conductor will be one-half the total watt load of two adjacent phases divided by the line-to-neutral voltage. In this case, 20 ranges, from Table 220-5, will have a total watt load of 35,000 watts for two phases; therefore, the current in the feeder conductor would be:

$$17{,}500 \div 120 = 146 \text{ amperes.}$$

On a 3-phase basis the load would be:

$$3 \times 17{,}500 = 52{,}500 \text{ watts}$$

and the current in each feeder conductor—

$$\frac{52{,}500}{208 \times 1.73} = 146 \text{ amperes}$$

Example No. 8. Motors, Conductors, and Overcurrent Protection

(See Sections 430-22, 430-24, 430-32, 430-52, 430-62, and Tables 430-150, and 430-152.)

Determine the conductor size, the motor-running overcurrent protection, the branch-circuit protection, and the feeder protection, for one 25-hp squirrel-cage induction motor (full-voltage starting, service factor 1.15, Code letter F), and two 30-hp wound-rotor induction motors (40°C rise), on a 460-volt, 3-phase, 60-hertz supply.

Conductor Loads

The full-load current of the 25-hp motor is 34 amperes (Table 430-150). A full-load current of 34 amperes × 1.25 = 42.5 amperes (Section 430-22). The full-load current of the 30-hp motor is 40 amperes (Table 430-150). A full-load current of 40 amperes × 1.25 = 50 amperes (Section 430-22).

The feeder ampacity will be 125 percent of 40 plus 40 plus 34, or 124 amperes (Section 430-24).

Overcurrent Protection

Running. The 25-hp motor, with full-load current of 34 amperes, must have running overcurrent protection of not over 42.5 amperes. The 30-hp motor with full-load current of 40 amperes must have running overcurrent protection of not over 50 amperes.

Branch Circuit. The branch circuit of the 25-hp motor must have branch-circuit overcurrent protection of not over 300 percent for a nontime-delay fuse (Table 430-152) or 3.00 × 34 = 102 amperes. The next larger size standard fuse is 110 amperes. (See Section 430-52.)

For the 30-hp motor the branch-circuit overcurrent protection is 150 percent (Table 430-152) or 1.50 × 40 = 60 amperes. Where the maximum value of overcurrent protection is not sufficient to start the motor the value for a nontime-delay fuse may be increased to 400 percent (Section 430-52 Exception (a)).

Feeder Circuit. The maximum rating of the feeder overcurrent protection device is based on the sum of the largest branch-circuit protective device (110-ampere fuse) plus the sum of the full-load currents of the other motors or 110 plus 40 plus 40 = 190 amperes. The nearest standard fuse which does not exceed this value is 175 amperes.

Example No. 9 Mobile Home

A mobile home is 70 × 10 feet and has two portable appliance circuits, a 1,000-watt 230-volt heater, a 200-watt 115-volt exhaust fan, a 400-watt 115-volt dishwasher and a 7,000-watt electric range.

Lighting and small appliance load

Lighting 70 × 10 × 3	2,100 watts	
Small appliance 1,500 × 2	3,000 watts	
	5,100 watts	
1st 3,000 watts at 100%		3,000
Remainder (5,100 − 3,000 = 2,100) at 35%		735
		3,735

$$\frac{3{,}735}{230} = 16 \text{ amperes per leg}$$

1,000 watt (heater) ÷ 230	4.4 amp
200 watt (fan) ÷ 115	1.7 amp
400 watt (dishwasher) ÷ 115	3.5 amp
7,000 watt (range) × .8 ÷ 230	24.0 amp

	Amperes per leg	
	A	*B*
Lighting and appliances	16	16
Heater (230 volt)	4	4
Fan (115 volt)	2	—
Dishwasher (115 volt)	—	4
Range	24	24
Totals	46	48

Based on the higher current calculated for either leg, use one 50-ampere supply cord.

ARTIFICIAL RESPIRATION

The mouth-to-mouth (or mouth-to-nose) method of artificial respiration is the most practical technique for emergency treatment of an individual of any age who has stopped breathing, in the absence of equipment or of help from a second person, regardless of the cause of cessation of breathing.

The mouth-to-mouth (or mouth-to-nose) technique has the advantage of providing pressure to inflate the victim's lungs immediately. It also enables the rescuer to obtain

ARTIFICIAL RESPIRATION

The mouth-to-mouth (or mouth-to-nose) method of artificial respiration is the most practical technique for emergency treatment of an individual of any age who has stopped breathing, in the absence of equipment or of help from a second person, regardless of the cause of cessation of breathing.

The mouth-to-mouth (or mouth-to-nose) technique has the advantage of providing pressure to inflate the victim's lungs immediately. It also enables the rescuer to obtain more accurate information on the volume, pressure, and timing needed to inflate the victim's lungs than are afforded by other methods.

When a person is unconscious and not breathing, the base of the tongue tends to press against and block the upper air passageway. The procedures described below should provide for an open air passageway when a single rescuer must perform artificial respiration.

Time is the most important single factor in successful artificial respiration. The human brain can exist for only a few minutes without oxygenated blood—probably not more than four or five. The greater the delay, the less chance of successful recovery. Do not waste valuable time by first loosening clothing, warming the victim, getting him down from a pole or moving him to a more comfortable position, etc. However, an immediate check of the victim's mouth should be made by a quick pass of the fingers through the mouth to pull the tongue forward and remove false teeth, chewing gum, tobacco, etc. After resuscitation is started, the victim's belt, collar and other clothing may be loosened, providing this does not interfere with resuscitation efforts.

The air you breathe out contains enough oxygen to save a person's life. If you breathe twice as deeply as usual, your exhaled breath contains enough oxygen for any adult victim. When each inflation expands the victim's chest, you can be sure rescue breathing is working. Inflate an adult's chest at least 10 times each minute. Infants require smaller and more frequent inflations—at least 20 times each minute.

Artificial respiration should be applied with a smooth, steady rhythm because a person normally breathes that way, but split-second timing is not necessary. However, the first half dozen or so cycles should be completed as rapidly as possible to correct the victim's oxygen deficiency quickly. The rhythm should then be slowed to the normal rate.

Time your efforts to keep in step with the victim's first attempt to breathe for himself.

If vomiting occurs, quickly turn the victim on his side, wipe out his mouth, and then reposition him.

Normally, recovery should be rapid, except in cases of electric shock, drug poisoning, or carbon monoxide poisoning. In these cases, nerves and muscles controlling the breathing system are paralyzed or deeply depressed, or the carbon monoxide has displaced oxygen in the blood stream over a period of time. When these cases are encountered, artificial respiration must often be carried on for long periods.

A brief return of normal breathing does not mean that resuscitation should be stopped. Frequently, after temporary recovery, the victim stops breathing again. He must be watched and if normal breathing stops, artificial respiration must be resumed at once.

Artificial respiration must be continued—four hours or longer, if necessary—until natural breathing is restored or rigor mortis sets in.

To avoid strain on the victim's heart when he revives, he should be kept lying down. It is extremely important that he be kept warm as well. If he revives before the doctor arrives, he should be given a stimulant, such as inhalation of ammonia, or a hot drink such as coffee or tea. A doctor's care is necessary during the recovery period, since respiratory or other complications may develop as an aftermath.

Relaxed tongue obstructs breathing

BEFORE STARTING RESCUE BREATHING, IT IS ESSENTIAL TO MOVE THE TONGUE FORWARD SO AIR CAN REACH THE LUNGS. Sometimes the victim who is not breathing will start breathing by himself if the tongue obstruction is removed.

During normal breathing, air flows easily through the nose or mouth to and from the lungs . . .

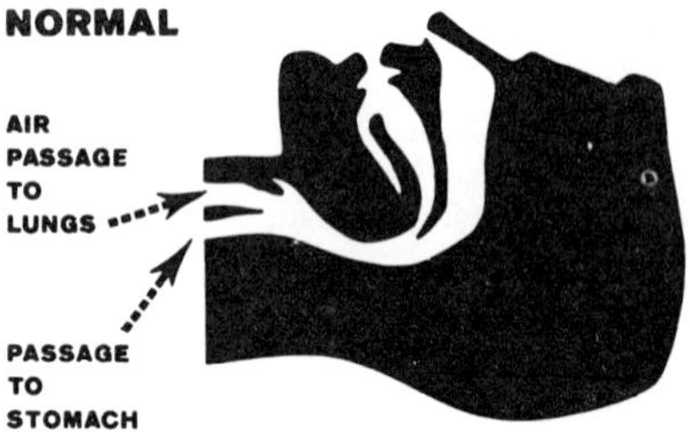

But when a person loses consciousness his relaxed tongue can completely block the movement of air through his throat . . .

How to prevent tongue obstruction

There are several accepted methods of keeping the victim's tongue from blocking his throat:

1. The **head tilt** is the most simple method.
   ... Using one hand, hold the crown of his head firmly and push backward.
   ... Pull his chin upward with the other hand.
   ... Hold his head tilted as far back as you can —until the skin over his throat is stretched tight.

2. **Lift chin** by grasping lower teeth with thumb.

3. **Lift jaw** upward, placing both hands on corners of jawbone near earlobes.

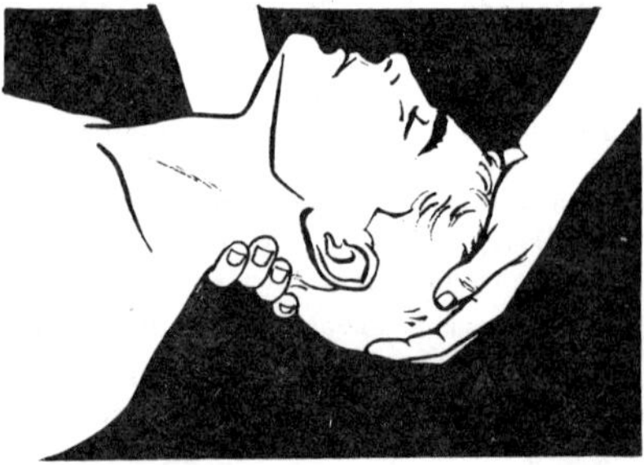

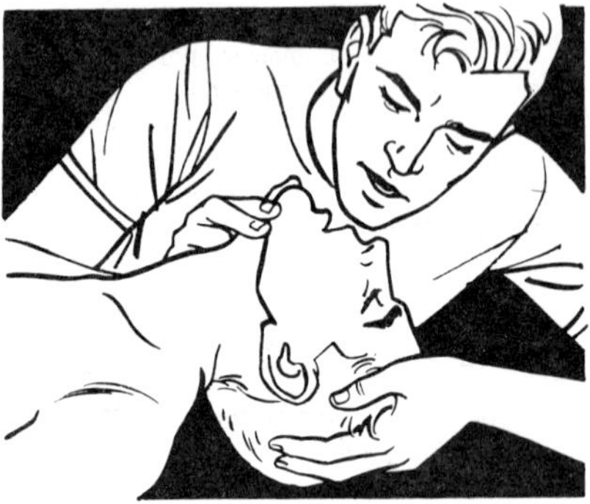

Rescue breathing for adult victims

Inflate chest 10 times **a minute**

STEPS TO FOLLOW:

1. If practical, place the victim on his back with his head lower than his chest. If not, leave him as he is. As long as you have access to his mouth or nose, rescue breathing can be done with the victim sitting in an automobile, pinned under debris, suspended on a safety belt on an electric power line, or floating face up in water.

2. **Lift** his neck and **tilt** his head back. This procedure is important to keep his relaxed tongue from blocking his throat.

3. Take a **deep** breath. Open your mouth **wide.** Blow air into the victim through his mouth or through his nose until you see his chest rise.

Mouth-to-mouth

Mouth-to-nose

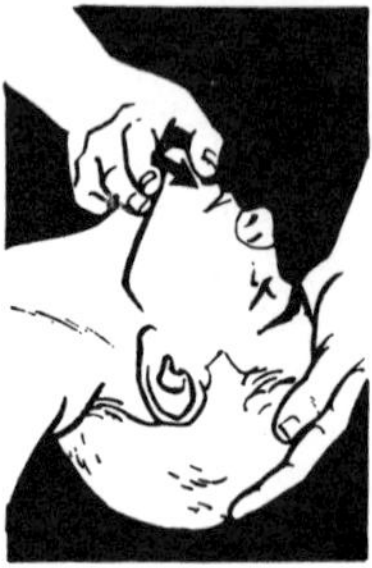

Mouth-to-mouth breathing — **Seal** your lips widely around the victim's mouth. With your thumb, **fold** his lower lip down to keep his mouth open during inflation and exhalation. To prevent leakage during inflation, **press** your cheek again his nostrils (or pinch his nostrils shut).

Mouth-to-nose breathing — **Seal** your lips widely on the victim's cheeks around his nose. Be sure your lips don't close his nostrils. **Close** his mouth with your thumb on his lower lip.

4. Remove your mouth and turn your head to one side, to let the air escape from the victim's lungs, to take your next breath, and to reduce the possibility of infection. When doing mouth-to-nose breathing, it may be necessary to hold the victim's mouth open so that he may also exhale through it.